CRYPTO

TRADING

BOTS

Crypto-Bot King

ABOUT THE AUTHOR

CRYPTO-BOT KING

CRYPTO-BOT KING are researchers

Based in London, England.

The CRYPTO-BOT King is a collective; we work with the most senior academic researchers, writers and knowledge makers.

We are in the changing lives business.

MASTERSHIP BOOKS

UK | USA | Canada | Ireland | Australia
India | New Zealand | South Africa | China

Mastership Books is part of the United Arts Publishing House group of companies based in London, England, UK.

First published by Mastership Books (London, UK), 2022

I S B N: 978-1-915002-12-9

Cover design by Rich © United Arts Publishing (UK)
Text and internal design by Rich © United Arts Publishing (UK)
Image credits reserved.
Colour separation by Spitting Image Design Studio
Printed and bound in Great Britain

National Publications Association of Britain
London, England, United Kingdom.
Paper design UAP

ISBN: 978-1-915002-12-9
- (paperback)

A723.5

Title: Crypto Trading Bots

Design, Bound & Printed:

London, England,
Great Britain.

CRYPTO

TRADING

BOTS

"An investment in knowledge pays the best interest"

- Benjamin Franklin.

CONTENTS

0

INTRODUCTION

Meaning of Cryptocurrency

Cryptocurrency is a type of digital money that is decentralized and based on blockchain technology. Although you may be aware of the most well-known versions, Bitcoin and Ethereum, there are over 5,000 distinct cryptocurrencies in circulation. They are a digital, encrypted, and decentralized medium of trade. Unlike the US dollar or the Euro, a cryptocurrency has no central body that oversees and maintains its value. Instead, these responsibilities are dispersed widely among Bitcoin users over the internet.

You may use cryptocurrency to buy conventional products and services. Still, most individuals invest in cryptocurrencies the same way they would in traditional assets like stocks or precious metals. While cryptocurrency is a new and exciting asset class, investing in it may be risky since you must conduct extensive studies to grasp how each system operates appropriately. Satoshi Nakamoto proposed the first cryptocurrency, Bitcoin, in a paper titled "Bitcoin: A Peer-to-Peer Electronic Cash System" in 2008. Nakamoto characterized the idea as "an electronic payment system based on cryptographic evidence rather than faith."

The Blockchain

A blockchain is a distributed, open ledger that stores transactions in code. In practice, it's similar to a checkbook that's spread among thousands of computers all over the world. Transactions are recorded in "blocks," which are then connected on a "chain" of prior Bitcoin transactions.

Each page is like a block, and the entire book, a collection of pages, is a blockchain. Everyone who utilizes a cryptocurrency has their copy of this book with a blockchain to build a uniform transaction record. Each new transaction is logged by software as it occurs. Every copy of the blockchain is updated instantaneously with the latest information, ensuring that all records are identical and correct.

To avoid fraud, each transaction is validated using two methods: proof of work or proof of stake.

Proof of Work

"Proof of work" is a way of authenticating transactions on a blockchain in which an algorithm generates a mathematical challenge that computers compete to solve.

Each participating computer, known as a "miner," solves a mathematical challenge that aids in verifying a set of transactions known as a block, which is subsequently added to the blockchain ledger. The first computer to complete the task gets awarded a deserving Bitcoin sum for its work.

The competition to solve blockchain riddles may need significant computing power and electricity. In reality, this implies that miners may barely break even with the cryptocurrency they earn for confirming transactions after deducting the expenses of electricity and computer resources.

Proof of Stake

Some cryptocurrencies utilize a proof of stake verification approach to decrease the amount of electricity required to review transactions. The number of transactions any user may verify using proof of stake is limited by the amount of Bitcoin they are prepared to "stake" or temporarily lock up in a community safe in exchange for the opportunity to participate in the process. "It's almost like bank collateral. Everyone who invests cryptocurrency is eligible to verify transactions; however, your chances of being picked grow with your stake.

Because proof of stake eliminates energy-intensive equation solving, it is far more efficient than proof of labor, enabling speedier transaction verification/confirmation.

If a stake owner (also known as a validator) is chosen to validate a fresh batch of transactions, they will be paid with Bitcoin, perhaps in the amount of the block's aggregate transaction fees. To deter fraud, if you are picked and verify incorrect transactions, you will lose a portion of the money you bet.

Consensus Mechanism

To verify transactions, both proof of stake and proof of work rely on consensus processes. While each utilizes users to verify transactions, each validated transaction must be reviewed and authorized by most ledger holders.

A hacker, for example, would be unable to modify the blockchain ledger unless they were successful in convincing at least 51 percent of the ledgers to reflect their fake version. The quantity of resources required to accomplish this makes fraud unlikely.

Cryptocurrency Mining

Mining is the process through which new cryptocurrency units are issued into the world, usually in return for confirming transactions. While it is theoretically feasible for a typical individual to mine cryptocurrency, it is becoming increasingly difficult to prove work systems such as Bitcoin.

As the Bitcoin network expands in complexity, more processing power is necessary. The typical customer used to accomplish this, but it's now too expensive. There are just too many individuals who have optimized their equipment and technologies to compete.

Also, keep in mind that proof-of-work cryptocurrencies demand a lot of energy to mine. It is believed that Bitcoin farms consume 0.21 percent of the world's electricity. That's almost the same amount of energy that Switzerland consumes in a year. It is believed that most Bitcoin miners spend 60 to 80 percent of their earnings on power.

While mining in a proof of work system is unfeasible for the typical individual, the proof of stake model requires less high-powered computation since validators are picked at random depending on the amount they stake. However, to join, you must already hold a cryptocurrency. (If you don't have any cryptocurrency, you have nothing to stake.)

Modern Cryptocurrency Applications

You can use cryptocurrencies to make purchases, but it is not yet a widely accepted means of payment. Although it is far from the standard, a few online merchants, such as Overstock.com, accept Bitcoin for payment.

You can work around existing constraints by trading cryptocurrencies for gift cards until cryptocurrency becomes more generally recognized. For example, at eGifter, you can use Bitcoin to purchase gift cards from Dunkin Donuts, Target, Apple, and a few more businesses and restaurants. You may put Bitcoin onto a debit card and use it to make transactions. You may sign up for the BitPay card, a debit card that converts crypto assets into dollars for purchase, in the United States. Still, there are costs associated with ordering the card and using it for ATM withdrawals.

You may also utilize cryptocurrency as an alternative investment to equities and bonds. The most well-known crypto, Bitcoin, is a safe, decentralized currency that has become a store of value like gold, a cryptocurrency specialist, and assistant editor for the financial news site Money Morning. Some people even call it 'digital gold.

Transaction Security

Using crypto to conduct safe purchases is dependent on what you're attempting to buy. If you want to spend Bitcoin at a store that does not take it directly, you can use a cryptocurrency debit card, such as BitPay, in the United States.

You'll need a cryptocurrency wallet if you want to pay someone or a store that accepts cryptocurrencies. A cryptocurrency wallet is a software application that interacts with the blockchain and allows users to transfer and receive money.

You can send money from your wallet by scanning the QR code of your receiver or by entering their wallet address. Some services simplify this process by allowing you to input a phone number or choose a contact from your phone. Keep in mind that transactions take time to complete because they must be authenticated using proof of work or proof of stake. This might take anything from 10 minutes to two hours, depending on the coin.

However, this lag period is part of what makes crypto transactions safe. A malicious actor attempting to change a transaction will lack the necessary software 'keys,' causing the network to reject the transaction. In addition, the network policies prohibit duplicate spending.

Making Cryptocurrency Investment

Peer-to-peer networks and cryptocurrency exchanges such as Coinbase and Bitfinex are places to buy cryptocurrency. However, keep a watch out for fees since some of these exchangers charge excessively high fees for minor crypto purchases. Coinbase, for example, charges 0.5 percent of your purchase price plus a fixed fee ranging from $0.99 to $2.99, depending on the amount of your transaction.

Recently, the investment platforms have begun giving the option to purchase some of the largest cryptocurrencies, including Bitcoin, Ethereum, and Dogecoin, without the fees charged by many significant exchanges. It used to be quite challenging, but it's now relatively simple, even for crypto virgins. "A non-technical exchange like Coinbase caters to them." It's effortless to open an account and link it to a bank account."

However, keep in mind that purchasing individual cryptocurrencies is similar to buying individual equities. Because you're investing all of your money into a single security, you're taking on greater risk than if you split it out among hundreds or thousands of securities, as you could with a mutual fund or exchange-traded fund (ETF). Unfortunately, crypto money is now scarce.

If you want to own a piece of the crypto economy, you might buy individual stocks in crypto firms. In terms of crypto-related equities, Coinbase is projected to go public sometime in 2021. "A few Bitcoin mining stocks, such as Hive Blockchain, are also available. If you want to have some crypto exposure with less risk, invest in large corporations implementing blockchain technology, such as IBM, Bank of America, and Microsoft.

Pros and Cons of Cryptocurrency

Before investing in and trading in cryptocurrency, there are certain key advantages and disadvantages to consider. This part will dispel many of your concerns, so let's take a closer look at both.

Pros of Cryptocurrency

Let's start with the good aspects of cryptocurrencies. Let's have a look at some of the benefits I've highlighted below.

- There is no fraud or scam in cryptocurrency.

There is no such thing as a scam in Bitcoin investment. Unlike conventional payment options such as internet banking, credit cards, or gift cards, Bitcoin payment can never decline. Payment is always delivered to the recipient in a matter of seconds. Because cryptocurrency is safe and digital, there can be no tampering.

- There is no risk of personal information leakage.

If you are purchasing something from a merchant shop and prefer to pay with your credit card, you should do so. You must offer your pin code to the seller to make a payment. What do you consider to be a safe method? Of course, a pin code is a highly personal item that you do not have to discuss with anyone. This is not a cryptocurrency issue, and you are not required to provide your private key to anyone. Even the money you make will be completely encrypted; no one will be able to verify your payment details, and it will never leak. As a result, it is the most incredible method for making anonymous payments.

- Secure and immediate ownership transfer

You have a valuable asset if you have cryptocurrencies in your digital wallet. You may quickly transmit it to anyone without obtaining permission. You'll need the other person's private key to complete the transfer. There is no price for transferring ownership, no stressful circumstance will occur, and no documentation is necessary. However, you must go through the extensive paperwork and pay the commission cost compared to other scenarios such as property transfer or bank account closing.

Cons of Cryptocurrency

We know you decided to invest in cryptocurrencies after reading the benefits, but it would be beneficial to be aware of the drawbacks.

- It is permissible to engage in illegal activities.

Typically, cryptocurrency is used to facilitate unlawful payments and activities on the internet. No government, however, can prohibit it since no one has power over it. Governments and top officials in the country can impose limits but not prohibit them.

- There is a high risk of loss.

The self-risk is a lack of ownership and power over cryptocurrencies. You have no recourse if something goes wrong. You cannot safeguard your Bitcoin against loss due to a technical issue in the wallet, and you cannot make a claim since no company controls this money. If you are experiencing any form of problem, you cannot report it to anybody. So, anytime you choose a wallet, make sure it has good evaluations and can be trusted. Another thing to consider before investing in cryptocurrencies is that you are aware of your country's current cryptocurrency regulations.

- The cryptocurrency market is quite volatile.

The Bitcoin market is quite volatile. Estimating the graph of Bitcoin value change is challenging, and it is impossible to anticipate when values will rise or fall. You must be a competent analyst and researcher and be up to speed on Bitcoin market news and trends. It would be best to prepare for any circumstance while investing in cryptocurrencies.

Takeaway

As you have seen, cryptocurrency has both benefits and drawbacks. Therefore, you must exercise caution while investing or trading in Bitcoin. Before investing in cryptocurrencies, consider all of the advantages and downsides. You cannot disregard the abovementioned advantages and disadvantages since they are significant and vital to your investment.

1

BASICS OF CRYPTO BOTS

Terms in Crypto Bots System

Access Point

A link to the exchange that enables TradeSanta software to execute the strategies you configure on the associated exchange. It would help if you first generate API keys on your cryptocurrency exchange to create an access point.

API Key

A one-of-a-kind identification generated by your exchange that enables you to trade using other applications

Dollar-Cost Averaging (DCA)

A trading technique in which the bot sets the first purchase order, followed by additional (buy) orders if the price moves in the opposite direction of the specified strategy. For all previous volumes purchased, just one take-profit order is made, and the take-profit price and volume are computed with each new purchase order.

Grid-Trading

This is a method in which the crypto bot sets the first purchase order and then buys more coins with further orders if the price falls, similar to DCA. The primary difference is that the Grid bot places a Sell order for each order rather than a single one for all purchases in DCA.

Current Market Price

This is the current rate of the cryptocurrency pair at the exchange you selected.

Balance

This is the coin deposit required for the initial order on your exchange account.

Dynamics

This is the percentage change in the selected trade pair's rate over 24 hours.

Volatility

Volatility is the percentage change in assets from the lowest to the highest price in 24 hours.

Take Profit (%)

This specifies the amount of profit you wish to make when selling the coins you purchased, expressed as a percentage of the order volume.

Step of Extra Order (%)

This specifies how much the cryptocurrency price should decrease for an extra order to be executed in the long and short strategy.

Max Extra Orders (%)

This is the maximum number of extra orders the DCA bot can execute in a single cycle. This parameter multiplies this quantity by the volume of extra orders. It adds the volume of the initial order to get the total amount the trading bot may invest in purchasing the currency.

Order Volume Number

Order volume number is the number of coins that the crypto trading bot will purchase at the start. The amount is specified in the quote currency (the pair's second currency) for the long strategy and the base currency for the short strategy.

Min Order Volume

The min order volume is the minimum number of coins that can be exchanged on the exchange. Usually, the equivalent of $11 suffices.

Max Number of Cycles

The max number of cycles is the maximum number of completed transactions, after which the crypto bot switches off and ceases trading until you restart it.

Martingale

Martingale raises the volume of each extra order by a coefficient ranging from 1.05 to 2 to recoup the value lost. It should be noted that this option is high risk and should only be used by experienced traders.

Max Order Price

The max order price is the price over which the trading bot will not begin a new cycle.

Stop Loss

A stop loss is activated when your current deal's loss reaches the value you choose in the options/settings.

Moving Average Convergence Divergence (MACD)

The trading bot will look for an appropriate entry point based on a trading pair's MACD during the last 100 minutes. The MACD indicator is used to confirm a trend.

Relative Strength Index (RSI)

The RSI is a momentum indicator that indicates whether the stock is oversold or overbought. TradeSanta's bot will discover an entry point for a trading pair based on the RSI for the last 100 minutes

Bollinger Signal

Santa's crypto bot will look for an ideal entry point based on a trading pair's Bollinger bands during the last 1.5 hours on 5 minutes.

Unsold Volume

The entire value of your position in base currency is known as the unsold volume. It is equal to the total of your existing First and Extra Orders for which no Sell orders have been executed.

Realized Profit

This is the profit you obtained after all sell orders were executed for the whole cycle.

Unrealized Profit/Loss (Liquidation Profit/Loss)

This indicates how much you will win or lose from your overall investment in a cycle. Unrealized profit/loss is computed as realized profit acquired during bot operation + (quote you will receive if you sell coins now - quote you paid initially for those coins at the time of acquisition) - trading fee.

Position

This is the number of coins you have in your possession that the trading bot is waiting to sell or buy back to earn a profit.

Long Strategy

A technique in which a crypto bot purchases a coin and sells it at a higher price to profit.

Short Strategy

This is a tactic in which a bot sells a coin and then buys it at a lower price to profit.

Take Profit Order

This order is set to profit by selling coins in the long strategy and purchasing them back in the short strategy.

Extra Order

The order is placed by the bot if the price moves against the bot's specified strategy (long or short). For example, if you have a long strategy and the coin price falls, it will acquire additional coins, and the cryptocurrency trading bot will sell more coins.

Force Close

This is a bot operation in which the bot sells positions and closes the current transaction. Profit/Loss will equal unrealized profit/loss in this situation.

Market Price

The market price is the current price at which the crypto asset is traded on the exchange.

Trading View

The trading view is a service that provides a set of instruments and indicators for doing technical analysis on many types of assets.

Introduction to Cryptocurrency Trading Bots

Trading Bot

A trading bot is a set of instructions delivered to a cryptocurrency exchange via software to perform transactions. Cryptocurrency trading bots are automatic trading systems that trade on investors' behalf. They enable you to make transactions automatically when certain conditions are satisfied. These bots consider information such as current pricing and volatility levels. In a nutshell, they simplify the investment process and make crypto-trading more accessible.

Bots are more efficient than people and make fewer mistakes, with less space for emotion or compassion. This is especially useful given the cryptocurrency market's often volatile price changes. Algorithmic trading bots are estimated to account for 70-80 percent of total crypto trading activity.

The majority of crypto trading bots provide the following services:

Data Analysis

Bots examine raw market data from many sources, analyze it, and decide whether to purchase or sell. Many bots allow users to customize the sorts of data they receive to deliver more detailed responses.

Risk Prediction

This is an essential element of a cryptocurrency trading bot. Bots use market data to predict an asset's potential risk, and this data assists the bot in determining how much to invest or trade.

Buying and Selling Crypto Assets

Crypto trading bots employ API keys (Application Program Interface) to buy and sell cryptocurrency assets strategically. The API key serves as a password for your trading bot to manage your account and place bitcoin orders. This is beneficial when you don't want to buy tokens in quantity.

Types of Bots

Crypto trading bots are highly customizable, allowing you to tailor the algorithm to your long- or short-term investing goals. Here are some of the most common types of crypto bots accessible today.

- **Arbitrage Bot**

Arbitrage Crypto trading bots, among the most popular, compare prices across several exchanges, and it then executes trades to profit on the price differences. Given the significant volatility of the cryptocurrency market, arbitrage bots assist in automatically setting Buy and Sell orders when there is a chance to profit. However, the gains from arbitrage bots might be modest.

- **Trend Trading Bot**

As the name implies, trend trading considers the momentum of a certain asset before placing a buy or sell order on it. If the trend indicates a price increase, the bot will open a long position. Similarly, a short position will be triggered if the price decreases.

- **Coin Lending Bot**

These bots allow you to lend coins to margin traders in the form of a loan that will be repaid with interest. Coin lending bots assist you in automating the procedure. You will spend less time looking for the best interest rate, and you will obtain better rates as a result.

However, it is crucial to note that crypto trading bots are merely instruments for automating trade. However, they may not be appropriate for everyone. It is critical to conduct a thorough study before selecting bots, and otherwise, you may wind up losing money.

Bots provide additional dangers. Programming flaws can also reduce the efficiency of these robots. Remember that bots operate based on the criteria and actions you specify. As a result, for a crypto trading bot to work in your favor, you must first have a solid grasp of cryptocurrencies and a solid investing plan.

Mechanism of Trading Bots

The most complex trading bots have three moving parts:

- Signal generator
- Risk allocation
- Execution

Signal Generator

This section is a prediction zone where crypto-trading bots forecast trade opportunities. The signal generator zone, which initiates the next stages, indicates the possibility of a buy or sell recommendation. These forecasting signals may be derived from current databases, use cases, and recent economic news.

Risk Distribution

Based on the generated indications, this section indicates the portion to perform the deal, whether buying or selling. These recommendations guide us to deploy cash for trading.

Execution

The last stage determines the value of your trade and executes bot predictions with your allotted cash to take action on your transaction. Each of the three components, signal, risk, and execution, requires its own set of algorithms and optimization methods. If you have a bot that fudges through any of these components, or worse, ignores them entirely, it will not serve you well in terms of profitability. To further understand how crypto trading bots function, consider discussing an example of a crypto trading bot. Bitsgap is one of the top crypto trading bots on the market, and it is utilized by many traders all over the world.

Using your API key, you may link your Bitsgap account to many cryptocurrency exchanges on the market. Bitsgap's trading bots can do an efficient market analysis and the actual trading operation.

Depending on the market conditions, you can utilize one of two setups. Assume that the market is gaining pace and that prices are expected to rise; in this situation, the Classic bot from Bitsgap is the best option.

As the price rises, the trading bot gradually sells the bitcoin to generate earnings. The bot can automatically balance risk exposure, which is fantastic. Another outstanding feature of this trading bot is its ability to generate consistent gains regardless of market circumstances. The trading bot can constantly locate even the tiniest possibilities to purchase cryptocurrency at a lower price and sell it at a higher price.

This guarantees that you can generate earnings steadily, which is critical in the cryptocurrency trading industry.

Crypto-Trading Bot Algorithms

- Momentum

If prices rise and we believe they will continue to rise, we will purchase. Profitability is minimal, but if the forecast is correct, the rewards will be rather high—the likelihood of winning is approximately 55%, and losing is around 70%.

- Mean-reversion

If prices fall and we believe they will rise in the future, we will sell. Profitability is high, and the gain-to-loss ratio is low—the likelihood of winning is approximately 70%, while the likelihood of losing is around 50%.

Trading Bot Vs. Human

- Longevity: Humans require sleep (and rest), whereas bots may function 24 hours a day, seven days a week.
- Speed: Robots are orders of magnitude quicker than humans in terms of thinking time plus reaction time.

- Emotionless: Greed or fear does not affect robots, and they will always do what is statistically more likely to result in a win.
- Capacity: Robots are capable of processing terabytes of data each second. Humans just cannot digest that much info in that amount of time.

You see, there are several advantages to using bots, and it all stems from their skillset being radically different from that of a person. Bots are monotonous and unchanging. To be lucrative, you must be consistent and, very honestly, do everything that goes against human nature.

However, it's important to remember that any bot will only ever be as good as the person who created it. The classic proverb "crap in, crap out" is correct. Humans can excel in certain areas, most notably with subjective reasoning.

A human is preferable when a specific piece of information does not have a definite consequence, and lateral or second-degree thinking is required to comprehend the meaning. I wouldn't be too concerned about this since when a bot reaches a pretty subjective condition, it can opt not to invest at all.

Factors to Consider When Selecting a Trading Bot

Before selecting a cryptocurrency trading bot, consider the following factors:

- **Security**

Effective security is critical for any online program, particularly crypto trading bots. If you use a bot created by an untrustworthy team with little or no online presence, you may end up putting yourself (and your money) in danger because they have access to your cash. Determining the security of crypto trading bots can be difficult, but look to internet groups and reviews to assist you in making the best, safest decision.

- **Dependability and Reliability**

Good crypto trading bots work around the clock and perform vital duties when you need them the most, which implies they must be dependable. If you select a bot that loses connectivity, you may lose a great opportunity to secure a profit even for a few minutes.

Look for reviews from other users that are as recent as possible. To acquire a range of perspectives from actual consumers, use some reviews or community sites. The more pleased people you can locate, the more trustworthy a bot.

- **Transparency**

One of the primary benefits of cryptocurrency trading is transparency: because networks are public, the possibility of unethical action and behavior is considerably minimized. A similar methodology is used by the safest, dependable, and recognized crypto trading bots.

Keep transparency in mind when looking for bitcoin trading bots and prefer trusted developers. An honest and transparent team will be more inclined to provide continuing assistance and advice.

- **User-Friendly Interface**

Because a bitcoin trading bot is intended to make the procedure as simple as possible, look for one with a user-friendly interface. A straightforward design and concise layout limit the possibility of being perplexed, and this is especially important if you're starting.

- **Profitability**

Do some study on the profitability of any bitcoin trading bots you're thinking about using. It's not worth utilizing one if you're not getting a good return on your investment. Check out what other people have to say once again.

Role of Bots in Crypto Trading

Cryptocurrencies are notorious for being extremely volatile, with values shifting substantially even within minutes. Investors may also trade bitcoins from anywhere in the world and at any time of day. When these characteristics are combined, they limit the effectiveness of human bitcoin trading in many ways.

To begin with, many investors are unable to react quickly enough to market movements to perform the finest transactions that are potentially available to them. Exchange slowdowns and transaction delays exacerbate the problem. Second, investors just do not have the time to commit to the cryptocurrency markets to make the best bets all of the time. To do this, bitcoin exchanges all around the globe would need to be watched 24 hours a day, seven days a week.

Fortunately, many investors can find solutions to these issues. One of the primary answers is bots or automated programs that conduct trades and execute deals on behalf of human investors. Bots are undeniably a contentious market component, and there are reasons to use them and reasons to shun them entirely.

- Traders use bots to profit from the global cryptocurrency markets, open 24 hours a day, seven days a week.
- Bots have an advantage over investors in that they can react more quickly.
- In the meantime, most investors do not have the time to commit always to obtain the best transaction, which bots can do.
- The arbitrage bot, which attempts to benefit from price disparities between exchanges, is an essential type of bot.

Benefits of Cryptocurrency Trading Bots

- **Absence of Emotions**

Because cryptocurrencies are so unpredictable, it's easy for traders to panic or overestimate their abilities. Emotionally driven decisions may result in judgment gaps and poor conclusions over time. However, one of the primary benefits of crypto trading bots is that they have no emotions or attachment to the money involved.

- **Efficient Trading**

Crypto trading bots excel at multitasking: they assess many currencies while also executing trades. The same narrow focus does not constrain them that individuals are, and they will not miss out on potentially profitable trades because they were focused on something else.

- **Faster Responses and Processing**

When the crypto market is volatile, taking too long to execute a transaction may be a major problem - you may miss out if you wait too long. However, because crypto trading bots issue quick orders, the possibility of missing out due to hesitancy is greatly reduced.

- **24 Hours Service**

After all, you're just human, and you will have to sleep at some point. On the other hand, Cryptocurrency trading bots operate 24 hours a day, seven days a week, making potentially profitable decisions for you while you sleep. They'll never pass up a good chance.

Takeaway

In conclusion, bitcoin trading bots may assist traders in achieving excellent outcomes, especially if you know how to modify them to your preferences. To maximize the value of your investment, however, it is vital that you read reviews, examine the creators, and confirm that you have chosen a secure option.

And to the question, do trading bots for cryptocurrencies truly work?

Yes, in a nutshell! Consider that Wall Street firms have been using algorithmic trading for many years. In truth, algorithmic trading bots have dominated the entire financial industry over the last decade, with algorithms now accounting for the bulk of trading activity on Wall Street.

So the question isn't whether they operate at all, but rather how effectively they work. And their effectiveness is determined by certain factors, including the platform and bots you choose and your level of expertise and experience.

On the other hand, Cryptocurrency trading bots are not a definite road to success, and they're partially automated, but not entirely. To trade efficiently, traders must grasp that building a quality bot involves specified goals, time, knowledge, and a certain degree of trust, which is why one-size-fits-all bots from unknown sources should be avoided.

2

CATEGORIES OF CRYPTO
BOTS PLATFORMS

Grid Trading Bot

A Grid Trading Bot is a trading bot that aids you in implementing the Grid Trading Strategy. It allows you to place a succession of buy and sell orders within a specified price range. Grid trading is a type of trading that entails placing a series of buy and sell orders at predetermined intervals around a set price. In this way, it creates a commerce grid.

When a sale transaction is completed, the bot immediately makes a purchase order at a lower grid level and vice versa. Grid trading is most effective in a competitive market with frequent price fluctuations. It ensures profitability if the selling price exceeds the purchasing price during a lateral price movement by automatically executing low purchase orders that lead to high sell orders, removing the requirement for market forecasting.

Several things must be addressed to increase grid trading earnings. What if I told you there is a way to profit from this volatility? What if you could automate your trades and simply sit back and watch your profits pour consistently?

A pair with constant and big ups and downs but a relatively stable long-term average price, for example, is almost likely a good candidate for Grid Trading. With everything out of the way, let's get down to business.

Advantages of Using a Grid Trading Bot

- **Reliable Crypto Trading Strategy**

Grid Trading has been around for a long time and is a well-established, tried-and-true, and lucrative trading strategy. There are several examples of successful traders using it for decades in different markets. Due to its great volatility, the crypto industry, in particular, has shown to be one of the most reliable venues for Grid Trading approaches.

- **Applicability/Usability**

The technique is simple to learn and use because it does not include complex calculations, measurements, or market indicators. It will be straightforward to set up for organizations that have no prior experience in the cryptocurrency trading markets.

- **Flexibility**

Grid Trading employs the most fundamental trading idea (buy low, sell high, earn the difference). Thus it can be applied to practically any market and create a profit regardless of trend or market behavior.

You may actively control the strategy's frequency and period by adjusting the price range and number of grids. Grids may be set up for the short term, capturing micro profit from all of the day's minor adjustments, or for the long term, by setting a big range and letting it run for months to benefit from every significant trend shift.

- **Improve Risk Management**

The ability to choose your grid approach lets you actively regulate the risk/reward level, which is not possible in most other kinds of trading. You may use a GRID Bot to produce a constant small profit with minimal risk (for example, by using a stablecoin pair such as BUSD/USDT), or you can take more significant risks for potentially enormous returns (e.g., with a low market cap coin that has high fluctuations)

- **Automated Trading**

Grid Trading is especially amenable to automation because of its obvious logic and that all of its operations are preset and independent of market behavior. Not only that, but employing Grid Trading with a Trading Bot is far easier and more efficient than manually following the method.

It's one of the best approaches for traders wanting to take their first steps toward automation since it can function on virtually any market, in any situation, 24 hours a day, 7 days a week, and for almost any duration.

Guides to Operate a Grid Bot

- Fees are an important issue since they have an ongoing impact on your trading outcomes. Exchanges with low fees that provide periodic zero-cost events or fee refunds as a market maker greatly impact your grid trading.
- Find the greatest pairs and market conditions you can and learn how to read the market, look at charts, and get a sense of what you need to look for.
- Look for charts with a sideways or slight rise, but make sure there are no signs of a long-term slump.
- Look for a pair that has previously experienced frequent and severe price swings but does not look to be poised to leave your price range very soon.

How it works

- As a limit order, the bot sets First Order (purchase order).

- Following the execution of the first order, the bot makes a take profit order and an extra order depending on the price execution of the first order.
- If the First Extra Order is executed, the bot takes the Take Profit order of the first order of the exchange, places the TP order for the executed extra order, and places the second extra order two steps below the First Order Price. The cycle is repeated when the second Extra Order is executed. There may only be one purchase and one sell order on the exchange simultaneously.
- The bot does not have deals as you are accustomed to with existing bots you utilize; instead, it operates indefinitely. When you terminate the bot, all active orders on the exchange are canceled, but the bot position (number of additional orders performed, quantity of coins purchased) is kept. All previously canceled orders will be replaced when you switch them back on.

Bot's Page

- The primary trade parameters that will help you track the bot's performance may be found on the right side of the page.
- Total traded volume is the total of all buy orders in the quotation currency that the bot has traded. Trades represent the number of buy and sell orders completed on the exchange (like trade history that you see on exchange). Please remember that the bot counts partially completed orders individually (each part is a trade). Assume you had a 50 USDT order volume, and the bot partly completed the first order for 30 USDT before buying 20 USDT. It will appear as two purchase deals in the Trades column, despite both trades being completed to match a single 50 USDT order at a certain price level.
- The bot progress table displays unsold volume, unrealized profit/loss, and current rate, as well as a graphic displaying where the most recent Buy and Sell orders were placed.

- The entire value of your position in base currency is the unsold volume, and it is equal to the total of your existing First and Extra Orders for which no Sell orders have been executed.
- Profit realized is the profit you obtained after all sell orders were executed for the whole cycle.
- Unrealized profit/loss (liquidation profit/loss) indicates how much you will win or lose from your overall investment in this cycle if you leave the position right now. Unrealized profit/loss is computed as realized profit acquired during bot operation + (quote you will receive if you sell coins now - quote you paid initially for those coins at the time of acquisition) - trading fee. Be mindful that if you have achieved profit but unrealized loss (i.e., the unrealized profit parameter is negative), you will have a loss if you quit the position right now.
- The Active Orders table displays orders that were placed but not performed. However, two orders can be displayed simultaneously (like open orders table on the exchange).
- Trade History displays all orders that have been fully or partially completed (i.e., like trade history you see on the exchange). Please keep in mind that trade is not the same as the order because if the order were completed in three parts, it would be displayed as three trades in this table.

Top Five Crypto Grid Trading Bots

- **KuCoin**

Kucoin is a Chinese cryptocurrency exchange that began it's operations in August 2017. It provides a comprehensive list of trade pairs. Over the past 24 hours, it traded in 432 markets, with a total trading volume of $25 million. It also has a comprehensive API to build a secure and automated trading strategy.

Fees: A 0.1 percent trading fee is charged to both the maker and the taker. The trading bot does not require a membership fee.

- **3commas**

3Commas is one of the best cryptocurrency trading bots, allowing you to increase profits while reducing losses and risks. You may benefit from our platform with minimum effort. It enables you to create a strategy based on over 20 trade indicators. Pricing starts at $14.5/month for the Starter license and goes up to $49.5/month for the Pro license.

- **Bitsgap**

Bitsgap is a leading bitcoin trading bot that lets you simply manage your cryptocurrency holdings. This program can assess over 10,000 cryptocurrency pairs and select the coin with the most potential. It enables you to create your bot strategy with only a few mouse clicks. Pricing starts at $19/month for the Starter license and goes up to $110/month for the Pro license.

- **Pionex**

Pionex is the world's first exchange, with 12 trading bots available for free. Users may automate their trades 24 hours a day, seven days a week, without checking the markets. It is one of the major Binance brokers, with Binance and Huobi Global liquidity.

Fees: A 0.05 percent trading fee is charged to both the maker and the taker. The trading bot does not require a membership fee.

- **Huobi Global**

Huobi, which has been in operation since 2013, is one of the largest cryptocurrency exchanges based in Asia and serving Asian consumers. Huobi Global provides an appealing platform for trading cryptocurrencies, with comprehensive support for several altcoins and a wide selection of stablecoins. Fees for both takers and manufacturers are 0.20 percent, and the trading bot does not require a membership fee.

Note

- If prices rise over your price range, the grid bot will have sold all of your positions, and you will be unable to profit from the higher price. It would have been wiser to buy and hold in this scenario.
- If prices fall below your range, it will utilize all of the cash to purchase the coin and incur losses when prices fall below your range. In this instance, even the finest grid bot in the world would struggle to generate a profit.

Best Grid Trading Bot Strategy

Grid Trading gains are divided into grid profits and floating returns. Grid profits will always be positive since they result from Buy Low and Sell High. When we execute Grid Trading, this is also our primary source of income. Floating returns can be positive and negative because they are determined by the price variations of the selected token. The floating return will be positive if you anticipate that the token price will progressively rise over time. This also shows us that using a token whose price has just dropped for grid trading is not a good idea.

Because all grid traders are focused more on grid profits, here are some strategies for increasing this portion of earnings:

- A sensible beginning price: You should enter the market at a modest price because a drop in the token price will result in floating losses.
- A suitable price range: If the price range is too large, your cash will be wasted, yet if it is too tiny, even slight price variations might cause the bot to exit the range.
- A sufficient number of grids: When the grid is too tight, the earnings in a single grid are relatively low, barely covering the trading cost in the worst-case scenario; yet, when the grid is set too broad, the money cannot be completely utilized.

Overall, a smart grid trader will create a sensible grid that optimizes money use based on historical volatility, funds, and trading expenses.

Takeaway

Grid trading is untouched by human emotions and is entirely governed by code. By analyzing market dynamics, grid trading will generate orders at gradually increasing and decreasing prices to profit from market volatility. As a result of the market's price discrepancy, individual investors will profit.

Martingale Strategy

Many people have probably heard of the "Martingale" strategy. This strategy has traditionally been used in speculative games, such as "High card takes it all." For example, if you lose one round of the game, you can double your bet in the next round until you win.

According to probability theory, if each round's winning rate is treated as an independent event, the final profit probability is 1–0.5n. That is to say, the more consecutive participation you take, the more likely it is that you will profit. For example, the first round has a 50% winning percentage, the second has a 75% winning rate, the third has an 87.5 percent winning rate, the fourth has a 93.75 percent winning rate, and the fifth has a 96.875 percent winning rate.

The chips you put in for the final win are twice as large as the previous one, so the previous loss can be offset by the winning one, resulting in a profit. For example, you put down $10 the first time, $20 the second time, $40 the third time, $80 the fourth time, and $160 the fifth time. If you win the fifth time, your total profit will be 160–80–40–20–10=$10.

Martingale is also referred to as a "Win with Enough Funds" method. If you have sufficient finances, you will undoubtedly be profitable in the end. But no one will have infinite funds. Even with a 99.99 percent winning percentage, one person may be unfortunate enough to lose the entire investment.

So, how should this martingale method be used in trading, and how can it be optimized to have a greater winning rate and higher capital utilization? Let's keep exploring.

Pionex Martingale Bot

Pionex Martingale Bot was developed with the typical martingale strategy's fundamental notion: a laddering-buy and the sell all at once strategy. And it will utilize more funds to buy on each downturn, dramatically lowering the average holding cost.

Pionex Martingale Bot will acquire uneven quantities of coins at a predetermined proportion following each price decline. When you establish a Pionex Martingale Bot, the bot will compute using the parameters you choose and split your investment evenly into multiple shares. The bot will then purchase coins with 1 share, 2 shares, 4 shares, 8 shares, 16 shares, and so on.

Assume that if the price falls by 1%, the bot will purchase some coins. As a result, the bot will purchase dips when the currency price falls to 99 percent, 98 percent, 97 percent, 96 percent, 95 percent. When it falls by 5%, the average price is 95.97% of the initial price. At this point, if the current price rises by 1.02 percent, it will repay the cost, considerably reducing the risk. Grid Trading Bot method, on the other hand, has to increase to around 99 percent of the starting price to repay its expenditures, whereas Pionex Martingale Bot just needs to rise 4.2 percent.

So, how can you generate money using Pionex Martingale Bot? The first step is to select nice coins. As long as this currency has excellent liquidity, the price will not fall to zero in the long run. The Martingale Bot can assist in earning gains; second, getting started is considerably easier. Most Pionex Martingale Bots will generate money unless the price continues to fall without any returns once they start.

Based on the preceding two principles, it is advised that users select the top coins with the largest market cap and begin at the Not-High price timing, which may generate returns with liquidity while reducing risks and depletion of your money.

Pionex Martingale Bot does not utilize leverage and allows you to specify the percentage of decrease for buying dips at your leisure. As a result, it is highly secure. Even if there is a short-term downturn, as long as the picked coins are good, they may be lucrative immediately as the price recovers.

Martingale Bot maintains its good product style — simple to use and appealing user experience. It has two approaches as well:

- AI strategy
- Manual configuration

AI strategy is divided into two types:

- Cautious
- Balanced

A cautious mind can minimize the strategy's maximum drawback, but the profit may be slightly reduced as well. Profits and risks will be more evenly distributed if they are balanced. If you're unsure, pick the balanced kind if you want to trade top coins with a high market cap, while the stable type is advised for volatile altcoins.

- **Artificial Intelligence (AI) Strategy**

Currently, the Martingale AI strategy is in its most basic form, with no settings dependent on the volatility of different coins. There are two sorts of AI parameters: balanced and conservative. The balanced type will purchase coins for every 1% drop and profit for every 1% earned, while the conservative type will buy coins for every 5% drop and profit for every 1% earned. The advantages and risks are modest for the balanced type, while the risks and returns are lower for the conservative kind.

- **Manual Configuration**

The manual configuration includes four settings, including common and advanced parameters. In the common parameters, there are three parameters: price scale, take profit ratio, and investment, with just one advanced setting: safety orders.

Pionex Martingale Bot operates in an automated cycle mode, which means that once you purchase dips and take profits, it will automatically join the following round. The following parameters are adjusted differently for each round.

Price Scale

After the initial order is executed, the bot will acquire coins when the price drops to this percentage in each round. This ratio can be filled with either a percentage or an absolute value.

- **Take Profit Ratio**

When buying drops and prices bounce to reach this ratio in each round, the bot will sell all coins to profit.

- **Investment**

This is the total money invested in this Pionex Martingale Bot.

- **Orders for Safety**

The bot will split the money evenly into numerous portions to purchase the dips in each round. Martingale Bot will buy dips using 1 share of the funds, 1 share of the funds, 2 shares of the funds, 4 shares of the funds, 8, 16, 32... to execute the orders.

The funds will be divided into 8 shares. Each round will buy 1 share (1/8 of the investment amount) at the start of each round, then buy 2 shares (1/4 of the investment amount) when the price drops the specific percentage, and buy 4 shares (1/2 of the investment amount) when the price drops the particular percentage. If the price falls again, the bot will no longer acquire additional coins and will instead wait for a price rebound that fulfills the take profit ratio before selling all eight shares to profit.

Pionex Martingale Bot Page

The Martingale information page is pretty clear and straightforward. The information page has eight parameters, and you may also use the information interface to see additional information.

First, the investment and total profit sections at the top of the information page display the total amount of funds invested and the total profit after subtracting the trading charge. As a result, the investment plus the entire profit will be collected when the bot is terminated. (Because of the period when the bot is paused, the market may vary significantly, causing some slippages. Therefore, the total amount received after the bot is stopped may differ somewhat from the one presented when the bot is stopped.)

The arbitrage profit, unrealized profit, arbitrage/total APR (Annual Percentage Rate), current price, take profit price of the coin and completed round number are presented below.

Arbitrage Profit

This is the overall profit gained after each round of profit-taking by Martingale Bot. Unrealized profit is the difference between the positive and negative earnings for coins that are yet to be sold in this round relative to the average cost.

Arbitrage/total APR (Annual Percentage Rate)

Arbitrage/total APR (Annual Percentage Rate) is defined as the arbitrage profits/total profit deduced into APR.

Current Price

The current price is the value of the coin with which you deal.

Take Profit Price

Take-profit price is the price at which the bot will sell all of the coins in profit in the current round based on the criteria you specify.

Rounds

Rounds are the number of times the bot assists in buying dips, and when the price recovers, the bot will sell the coins to profit, which is a full circle as one round.

Grid Trading Bot vs. Pionex Martingale Bot

The main difference between the Martingale Bot and the Grid Trading Bot is that the Martingale Bot will ladder both purchase and sell simultaneously, whereas the Grid Trading Bot will ladder both buy and sell. Because Martingale Bot will buy more and more coins as the price falls, it will maintain a tiny position with the coins when it initially starts. In contrast, Grid Trading Bot will start with more coins based on the settings supplied by the user. Typically, the coin will absorb around half of the initial investment. As a result, when the market rises, Grid Trading Bot gets larger trend gains than Martingale, but when the market falls, Grid Trading Bot suffers a bigger loss.

On the other hand, Martingale Bot splits funds into 32 shares by default. Furthermore, Grid Trading Bot frequently divides money into more than 100 shares to chase more frequent arbitrage profit. Therefore, Martingale Bot's capital usage rate will be higher than Grid Trading Bot's. When the market is upswings, Martingale Bot will take a more significant arbitrage profit than Grid Trading bot.

So, to summarize, the benefits and drawbacks of Martingale Bot and Grid trading are as follows:

- Trending Profit Martingale Bot < Grid Bot
- Risk Martingale Bot < Grid Bot

- Arbitrage Profit Martingale Bot > Grid Bot

Rebalancing Bot

Need for Rebalancing Bot

You can use the rebalancing bot if you are positive about many coins simultaneously and are prepared to retain coins for a long period to earn value appreciation.

Rebalancing Bot's Trading Strategy

If you have 100U and opt to build the rebalancing bot using BTC and ETH, the bot will distribute your position at a 1:1 ratio. The bot will purchase 50U of BTC and ETH based on the current BTC and ETH values. By default, the bot will monitor every 5 minutes and update the position based on the exchange rate variations between the two cryptocurrencies, ensuring that the position ratio is always 1:1.

When the value of your BTC has climbed to 60USDT, but the value of ETH has remained stable at 50USDT, the rebalancing bot will sell 5USDT of BTC and purchase 5USDT of ETH to maintain the position ratio at 1: 1. At the moment, your BTC position is 55USDT, and your ETH position is 55USDT.

Compared to retaining the coins, the rebalancing bot has the benefit of being able to configure the bot's rebalance criteria and leverage the exchange rate changes between the tokens to carry out rebalance arbitrage to accomplish the goal of earning coins.

How to Make a Rebalancing Bot

Select a bot

To begin, go to pionex.com and log in to your account, then click the [CREATE] button on the right side of the rebalancing bot. There will be two modes displayed on the page: [Dual-coin mode] and [Multi-coin mode].

Dual-coin mode

You may allocate two coins at the same time in "Dual-coin mode," and it supports "Spot" and "Leveraged" coin selection. It is advised that everyone selects major coins of spots, like BTC or ETH. However, because leveraged tokens are high-risk derivatives, please be cautious while picking them.

When you've decided on the two currencies, you may choose the amount of money that will be invested to begin the rebalancing bot.

By default, the position allocation ratio of the two coins in "Dual coins mode" is 1:1. When the exchange rate between these two currencies varies, the bot will automatically restore the two coins' real-time position ratio to 1:1.

Multi-Coin mode

You may allocate up to 10 coins in "Multi-coin mode," and it supports "Spot" and "Leveraged" coin selection. Each coin's position proportion may be arbitrarily adjusted. If you're unsure which currencies to use to develop the rebalancing bot, you may start by clicking the "Choose an index" option.

You can select the index based on your preferences.

The advanced setting area is located below the manual setting page. We offer numerous criteria from which to pick.

- Trigger price: The trigger price for one or more coins can be specified. The rebalancing bot is formed and activated when a coin hits its trigger price.
- Rebalancing: You have the option of turning off rebalance (the bot stops operating, holding the currency, and remaining unaltered) or turning it on (the bot starts to run and keeps the proportion of each coin in constant through rebalance).

You may select between two rebalance modes when you enable

rebalancing: periodic and threshold. When the coin's price fluctuates, the value of each coin deviates from the predetermined ratio. At this point, the assets must be rebalanced to return the percentage of each currency to its original condition.

When selecting "Periodic," the bot will rebalance the asset ratio at predetermined intervals. When you select "Threshold," the bot will rebalance the assets if the proportion change of any currency is larger than the stated threshold.

3

GETTING STARTED WITH BOTS

The Components of a Crypto Trading Bot

All trading bots have the following components in common:

- Strategy Implementation
- Backtesting
- Job scheduler
- Execution.

Strategy Implementation

It is vital to plan out the trading strategy that your bot will use. At this stage, you define the logic and computations that will help you determine what and when to trade. After developing the approach, you must test it to determine how well it works. In a moment, we'll go through several tactics you can use.

Backtesting

You must backtest your bots against previous market data before making any transactions with them, and you should make your backtest as realistic as feasible. You can do so by considering the latency, slippage, and trading costs. Accessing exchange APIs allows you to gather high-quality market data, and libraries like CCXT can help you interact with many transactions.

Job Scheduler

Now that you've hard-coded the techniques and put them through their paces in the real world, it's time to automate the entire process. You must set up a task scheduler to execute your trading techniques automatically.

Execution

After you backtest your strategy, the next stage is to put it into action in real-time. The logic that you hardcoded into the bot will now be transformed into API calls that the exchange can comprehend. Some bots may have even let you test your technique in real-time with fictitious money.

Creating Cryptocurrency Trading Bot

The ability to retain control over your private keys is the most obvious advantage of utilizing an independently modified trading bot. You may also add any functionality you like to the trading bot. Furthermore, once the trading bot is configured, you may trade 24 hours a day, seven days a week, increasing your chances of profit through speedier order placement.

The cryptocurrency market is rising daily, as is the number of trading bots. Most advanced crypto-trading bots are currently rather expensive to purchase or are available on a subscription basis. Nonetheless, there is now a more natural approach to obtaining a trading bot. Free trading bot software is available on many open-source platforms for anybody to use. 3Commas is a well-known example. 3Commas provides a trading bot that works well with different exchanges, including Bitfinex, Binance, Bitstamp, GDAX, Huobi, etc. With a patched-up 3Commas trading bot, one may trade 24 hours a day, seven days a week, and even mimic successful traders' techniques.

Creating an API for Each Exchange

APIs have become increasingly crucial in today's bitcoin trading ecosystem, and most consumers are unaware that they are utilizing one. An API (Application Programming Interface) is a trading bot interface that enables the bot to send and receive data from an exchange. Most cryptocurrency exchanges let you utilize their API access for the bot. These systems, however, are often predicated on a few permission levels that are protected by unique keys and secrets. To do so, you must first create an API key on the exchange you intend to utilize. After that, you'll need to enter your key and secret into the trading bot to access the API.

API keys are essential. It is recommended that you never discuss your API secret keys with anybody and that you use extra caution when choosing which platform to put your key into. If the keys are lost or hijacked, someone else can get access to your trading bot and use it to trade or make withdrawals without your knowledge. If the API has the withdrawal option enabled, you should probably disable it. Turning it off prohibits the bot from removing funds from your account and allows you to make manual withdrawals.

API for the best currency exchanges:

- Bitstamp
- Bittrex
- Kraken
- Poloniex
- BitMEX
- Gate.io
- Binance

Create a Checklist for Bitcoin Trading Bots

Instead of paying for a subscription or purchasing one, you may create your trading bot. Here are some checklist steps you may do to ensure you build an intelligent trading bot with little hassle.

- Choose the programming language you wish to use. It is a good idea to select a standard or known programming language so that you can easily bring in development help if necessary.
- Take control of your APIs. Ensure you have all of the APIs for the crypto exchanges with which you want your trading bot to interface. For example, if you're building a GDAX trading bot, you'll need access to the GDAX API.
- Set up accounts with the exchanges you wish to utilize. For example, if you want to build a Bittrex trading bot, you'll need access to the Bittrex API.
- Select a trading bot strategy, whether it's market creation, market following, or arbitrage. The longer it takes to build a trading strategy, the more complex it is.
- Architecture. Make sure you describe what kind of data you want your trading bot to comprehend.
- Create. The most time-consuming aspect of developing the trading bot is its creation. Make sure that you follow all of the proper procedures.
- Test. Check that your trading bot is working correctly. If it isn't, now is the moment to fine-tune it.
- Deployment. Once you've resolved any difficulties with the trading bot, it's time to deploy and utilize it.

Investing in Creating Trading Bot

You will need to invest in a specialized automated trading program to become a full-fledged market maker and engage in professional bitcoin trading. Even if some users pick free or paid bots from third-party developers or even open-source projects from Github, the consequence of their activity may be inconsistent, and their performance – questionable.

A custom solution is required to leverage computer algorithms for cryptocurrency trading with optimal efficiency. Is it on the expensive side? Yes. Is it worth your time and money? Definitely! Custom software is the sort of investment that pays for itself immediately while also saving you money and anxiety in the long term.

Cash Requirement

Your time is important, so let's get straight to the point: there is no set pricing for developing a crypto trading bot. Given the staggering amount of variables impacting the ultimate pricing, there cannot be. It contains the number of needed features or the period and the number of developers, their level, post-dev support, the location of the contractor, and much more.

Five Elements that Influence the Creation of Trading Bot

- The number of functions implemented

There is a clear correlation: more features necessitate more time, personnel, and money. To reduce resources, you may, for example, omit reports, a scheduler, and market tracking from the functionality.

- Platforms and currencies that are supported

You might try to cover as many platforms as possible or just a few transactions. Alternatively, you might concentrate your efforts on a single, promising cryptocurrency exchange, such as Binance, Bittrex, Bitstamp, Coinbase, and so on. Similarly, you can limit the coins supported by the automated trading program. As a result, developing a bitcoin trading bot will be easier and less expensive than developing sophisticated software supporting many cryptocurrencies.

- Strategies and customization options

Again, your program may be focused on a particular technique; however, it will have far greater potential if it enables a variety of cryptocurrency-related activities. These actions determine the kind of trading bot. Furthermore, a well-designed trading bot should enable user strategy creation, testing, and execution. Using this feature, a user may manually fine-tune the bot's behavior by altering strategy parameters, circumstances, and rules.

- Improved quality of life

These are small functions that may be easily eliminated to save money, and they have no direct influence on the bot's functionality and boost user convenience. Optional features may include, for example, a notification function and a clean UI design.

- The technology stack and the availability of specialists

Because these two aspects are inextricably linked, it is best to treat them as one. Existing bitcoin trading bots are typically (but not always) developed in Python, PHP, or Javascript (specifically Node.js). Therefore, you'll need a team of competent engineers well-versed in these languages.

If you are having difficulty finding a great fit or if your budget is too limited for the local IT market, bear in mind that you always have the alternative of outsourcing. For example, the hourly compensation of web developers in Ukraine is about twice as cheap as in the United States or Western Europe, despite the consistently good quality.

Strategies for Cryptocurrency Trading Bots

Most trading bots in the bitcoin market employ a few trading methods by default; nevertheless, these tactics are not as complex as most people believe. Rather, they are basic algorithms that take advantage of the bitcoin market's tremendous volatility.

Today's crypto bot methods necessitate fast order placement, which humans cannot always do. They also necessitate a quick computation of figures, so they are important while trading in the bitcoin market.

- Machine Learning/Artificial Intelligence

Machine learning is still in its infancy in the crypto trading world, and it employs artificial intelligence to analyze historical data and translate it into a viable trading strategy. Essentially, the machine-learning algorithm learns from the data given into it and improves its ability to predict the price of a certain bitcoin asset.

However, this trading approach is not as straightforward as it may appear. Artificial intelligence learning is a hard process requiring a large amount of data. Furthermore, the machine-learning algorithm cannot consider crucial news, market mood, and other essential elements.

- Grid Trading

This cryptocurrency trading method employs stop buy and stop sell orders to execute a large number of minor deals. It also establishes a stop loss and a profit target in advance. However, one of its most essential advantages is that it can assist traders in predicting whether the bitcoin market will move up or down. The downside of this method is that it increases the possibility of human mistakes. This is because traders must manage many deals at the same time. This is when cryptocurrency trading bot tactics might come in handy. They can manage a large number of deals at the same time.

- Signal Trading

This trading technique employs several technical indicators to determine a solid entry position. Among these indicators are EMAs, MACD, RSI, and others. The RSI is likely the most widely used indicator. Essentially, it examines the market's motion. When it exceeds 70, it often indicates that a certain digital currency is overbought, indicating impending pullback. However, when the figure falls below 30, it indicates that the cryptocurrency has been oversold, suggesting that a bounce is imminent.

Exponential moving averages (EMAs) are not the same as the Relative

Strength Index (RSI). Moving average convergence divergence is often generated by using at least two EMAs. Furthermore, they can employ many technical indicators in tandem and only join trades when favorable conditions occur. Because several technical signs might contradict each other, this can be a complicated method to employ.

- Ping-Pong

This is one of the most straightforward bitcoin trading bot tactics. As the name implies, this is a buy low, sell high trading technique. To use this method, traders must place a purchase order followed by a sell order as soon as possible, with a larger proportion of profit. When the sell order is filled, traders must immediately place a purchase order at a lesser proportion.

They can keep playing ping-pong for as long as they wish. This is an excellent method for trading bots, and they can rapidly establish all orders and determine any proportion traders desire to place. Furthermore, traders can leave the trading bots running indefinitely.

- Bollinger Bands

Traders can utilize Bollinger bands independently; however, they are often used with other technical indicators. They are narrow when the market is trading sideways and broad when the market is turbulent. This trading approach is quite effective in the bitcoin market. The market will be trading sideways to see whether a huge move is on the way. This is not frequently the case in normal stocks. This is because equities can move sideways for numerous years.

- Breakout Strategy

This trading method employs a defined level, which many traders employ to increase the likelihood of a large continuation and a successful entrance. It uses the motion of a coin and is pretty simple to grasp. Traders must first identify support and resistance levels before selling or buying orders.

Putting your purchase order just above the resistance level and your sell order slightly below the support level is critical. Stop losses are the sole safeguard against fakeouts. As a result, traders employing this method should bear this in mind and comprehend how the crypto bot strategies now in use might assist them.

How to Select Trading Bots

Crypto trading bots come in a variety of forms and designs. Several platforms are available, each with its own set of features that allow you to automate your trading experience entirely.

Here is a list of factors to keep in mind when looking for the finest platform.

Team Reputation

Above all, the team's reputation is important. Hackers have crippled unwary users by exploiting ingenious flaws. One example is when hackers utilized trading APIs to inflate the value of Syscoin on Binance artificially. Because hackers are astute, having a trustworthy founding team is the first line of security.

When examining the team, look for characteristics that indicate a respected team. This covers things like:

- Where the players of the team went to college.
- What businesses the members worked for.
- The length of time the team has been working on the bot.
- Members of their advisory board's reputation.
- The startup's financing source.
- Incorporation country.
- The original team's technological abilities.

Before we put our faith in a third-party development team to automate our strategies, we should utilize this list to learn more about the team behind the service.

In addition to the team's particular members, ensure that the organization may be held responsible for any possible misconduct. To know if we can trust the team, it must be clear about its origin, its members, its base, and any other pertinent information.

Automated Strategies

We need to ask if the bot we're looking at genuinely implements the strategy we want to automate. If we wish to rebalance our portfolio, using a trading bot that only supports techniques involving moving average cross-over, for example, is ineffective.

Examine the trading bot's website to ensure that it supports the tactics we wish to use for our portfolio. Examine any configurations they provide or clues at the amount of participation required from an end-user. It's also pointless to use a bot if it helps our approach but ends up needing more effort than having us apply it.

Support

The degree of assistance supplied by the team is the second most significant factor to evaluate. Is there a way to contact the development team if you have a support question or need assistance with a bug? If not, you may be left hanging for weeks without answers to essential concerns about your plan. We have witnessed this several times and can guarantee you that it is never a pleasant experience.

The following are some ways we may tell if the team delivers excellent support:

- There is a lively Telegram or Discord group where the team can address queries.
- Contacting the team is made easier by the availability of many channels.
- The support crew responds quickly to inquiries and actively attempts to address issues.
- Their conversation networks, such as Telegram and Discord,

receive frequent updates.

These are just a few methods to gauge the kind of help you will receive if you decide to sign up.

Cost

Almost all services and goods in the blockchain business are expensive, and trading bots are no exception. While firms provide affordable and well-budgeted price packages, solutions still cost thousands of dollars.

Keep your budget in mind while selecting a platform. Consider crypto trading bots to be tools that can assist you rather than things that may completely turn your life upside down. If you are new to the cryptocurrency market, investing your money in expanding your portfolio is better than purchasing pricey trading services.

Ease of Use

The fact of the bitcoin industry is that most products are designed in an unduly sophisticated manner. This unquestionably applies to trading bots as well. The majority of trading bots are quite complex, and they have an enormous number of levers, choices, setups, and settings. Diving immediately into the frigid pool's deep end might take your breath away and fill you with fear.

Look for a bot that corresponds to your degree of comfort. Instead of diving headfirst into open-source trading bots and attempting to create your techniques by exercising your development talents, perhaps start slowly. Experiment with a bot that simply offers a few key capabilities but is simple to operate. That way, you may put your toes into the pool and have a feel for it before diving in.

Backtesting

Backtesting is the process of simulating a strategy's performance through past data. It assists us in understanding the strategy's behavior under particular situations and can aid us in determining how the approach will perform in the future. Of course, nothing is certain, but this is the most rigorous method for assessing tactics.

Backtesting should be included in the bot we pick. We never propose using a technique until it has been thoroughly tested. Going into the future blindly might lead to calamity. Spend some time evaluating the approach, determining the ideal configurations, and implementing the plan in a way that corresponds with your goals, as well as the knowledge gleaned from the backtesting tool.

Caution! Any backtesting program that uses CoinMarketCap data should be avoided, and this data source cannot be used to evaluate trading strategies. This data is grossly wrong, but CoinMarketCap combines data from several exchanges. Therefore, it is not true bid-ask data from exchanges. Backtesting techniques should only employ accurate bid-ask data acquired directly from each individual exchange.

Execution & Implementation

Evaluate the strategy's implementation carefully to assess how the approach will perform under different scenarios. Because each bot has a distinct strategy implementation, it's critical to grasp the variations between each trading bot's implementation. Not all implementations are regarded in the same way. Consider the bot's available controls, setups, and the strategy's resilience under changing situations.

At the same time, just because the trading bot includes the essential restrictions for implementing the strategy does not imply we are safe. Everything might appear promising, yet it could all fall apart due to bad execution. The programming and infrastructure created around the trading bot are critical to handling the strategies. Failures caused by server difficulties can result in improper trades, mistiming, and overall poor strategy execution.

Losing money due to poor plan execution may rapidly turn into a nightmare, and you will not be able to recover your cash. Therefore, avoid the problem by picking a trading bot that is being developed utilizing acceptable software practices. Examine the team's willingness to address issues, join their marketing channels to check the frequency of bug reports, and keep track of how long it takes the team to handle the issues.

This brings the current lesson to a close. The next course demonstrates how to create your own crypto trading bot and use most platforms' extensive scripting capabilities.

4

THE LEGALITY
OF TRADING BOTS

Different regulatory frameworks allow traders to perform more or less acceptable things. So, when it comes to cryptocurrency bot trading, the basic answer is:

Yes, trading using crypto bots is entirely legal. There are, however, certain drawbacks. Just because it is permitted to deal with bots does not imply that anything involving bot trading is acceptable. Driving a car is legal, but there are many things you can't do without. Some platforms prohibit traders from using bots for many practical and regulatory reasons.

More than just your trade is involved in automatic trading. There is also the issue of bot supply. Some items are lawful to use but not to purchase or sell. It makes a difference whether you are using a bot you made, whether the bot is available for free download, or whether someone is earning from a bot with particular advertised qualities.

Bots are Not the Same as Traditional Trading

As previously said, there are several types of trading bots. The distinctions between bot types have a practical impact on how you engage with the market. As a result, regulators' perceptions of what you're doing alter, as does their willingness to accept it.

Semi-automatic bots can be ruled out in terms of regulation. These systems monitor the market and do more complex calculations to propose a transaction to you. However, in the end, you must determine whether or not to engage in the transaction yourself. This is manual trading, aided by a program/bot for practical purposes.

The real problem arises when fully automated trading bots gain access to your account and trade the market without your direct participation. That's where the majority of the caveats come into play. This is due, in part, to the fact that they are much riskier (their selling pitch is that you have no control over what they do with your money), and they are frequently sold by organizations looking to benefit from selling bots rather than creating gains in the market.

Being cautious does not need to be mandated by law, and bots are far from certain to generate profits. Even the greatest ones adhere to set rules and cannot adapt to market developments. On the other hand, people selling bots often aim to persuade potential consumers that their bot is superior and worth purchasing. As a result, even truthful merchants may inflate the benefits of their bot to absurd proportions, and that would be against the law.

However, many people are merely scamming by selling bots that do not operate as stated. That is much more unlawful! Many platforms do not accept bots, not because they generally dislike bots, but because they are concerned that a trader may purchase a bot that will damage their account. This is why, if you ask most experienced traders, they will tell you not to use bots until you've mastered manual trading. This will assist you in identifying frauds.

The legal issue with employing what is commonly referred to as a "scambot" is that it is lawful for you to utilize it and lose money. However, because the seller is in another jurisdiction, they cannot be held legally liable for misleading advertising and the loss of your money. You must exercise extreme caution when it comes to claims of fast money. And just because something is legal does not guarantee it is safe, beneficial, or even desirable!

Bots Aren't Magic; They're Tools.

Algorithms are increasingly being employed in cryptography. Furthermore, automatic trading might help you increase your profits. They are, however, instruments to help you with your trade. They should not be viewed as a replacement for your decision-making abilities. And especially not as a substitute for getting market experience and knowledge!

Reliable platforms will go out of their way to educate their clients on how to trade, so before utilizing bots, read what your platform has to say about them.

One of the issues with trading bots is that the marketplace is rife with scammers. As a result, their validity is frequently called into doubt. There are dozens of examples on the internet of firms and people selling bots but failing to provide the results promised. The majority of victims of these frauds do not retrieve their money. The vendors frequently offer money-back guarantees, but they seldom respect these agreements, and in many cases, they close before consumers realize what is going on.

A few businesses offer actual bots with some promise away from the frauds. The only method to ensure a reliable trading bot is to build one yourself. If you have a tried-and-true trading strategy that adheres to a set of rules, you can turn it into a software. Of course, this necessitates familiarity with the programming language used by your favorite trading platform.

If you cannot create the bot, you may hire a specialist you can trust not to resell your method and perhaps render it ineffective.

The thought of a bot trading on your behalf may appear suspicious, yet the practice is generally encouraged. According to CryptoCurrency Facts, because a thin market is terrible for everyone, the more buy and sell orders on the books, the better. Hence bots are absolutely encouraged. Bot trading is common in both the cryptocurrency and stock markets, although not all brokers allow it. Furthermore, everything prohibited in traditional trading environments is likewise banned in crypto trading. This implies you can't precisely construct or purchase a spoof bot without generating problems; instead, you should emphasize making or picking a bot that follows indicators.

5

EARNING MONEY
ON THE PLATFORMS

Methods of Making Money in Cryptocurrency

There are several methods to profit from cryptocurrency trading: Crypto trading is comparable to stock trading in that we may purchase and sell tokens/coins for profit margins. The volatility of cryptocurrency is enormous, as is the possibility of profit.

Purchasing and HODLing

HODLing is the process of purchasing powerful cryptocurrencies with outstanding concepts and solid use cases and then holding them for an extended period with the expectation that their prices will rise dramatically. When the coin price has risen sufficiently for you to benefit, you may sell it on the cryptocurrency market.

Cryptocurrencies such as Bitcoin, Ethereum, Ripple, Cardano, Litecoin, BNB, and others are ideal for long-term investment and HODLing. HODLing, on the other hand, does not yield you any interest in your money, and it assists you in making the transition from buy-and-hold to active trading.

A large proportion of cryptocurrency investors uses the buy-and-hold approach. These individuals, who are commonly referred to as "crypto investors" rather than "crypto traders," gain money by relying on the long-term development in the value of a currency.

People choose this passive method for many reasons, the most prevalent of which is a lack of time and understanding. Being an active trader necessitates a constant focus on the turbulent markets of cryptocurrency. This is hard for many people who have day jobs outside of the crypto world.

While there is nothing inherently wrong with the buy-and-hold approach, the truth remains that active trading provides more profit potential than passive investment. If you can regularly execute winning trades the majority of the time, you'll make far more money than someone who prefers to purchase and hold.

What is the solution? That's right, a cryptocurrency trading bot!

Earning Money with Crypto Trading Bot

A cryptocurrency trading bot assists you in capturing as many cryptocurrency upswings as possible. It attempts to minimize your exposure to downswings, all while operating independently. This means you don't have to spend all day staring at a computer to time the market, and the bot does everything automatically.

You may go about your everyday activities while the crypto bot executes transactions for you. So, if you're searching for a simple way to convert from passive investment to active trading, this product is well worth a look.

Profitable Methods

Staking

Staking is the process of investing or locking up cash in a cryptocurrency to earn new cryptocurrencies in the form of interest.

Furthermore, if you choose to retain your coins for a specified amount of time, you will profit from price appreciation.

Bonus Coins/Tokens

Participating in bonus/airdrop offers is another excellent method to make money with cryptocurrencies that don't involve any investment. To get some free/bonus coins, all you have to do is join a cryptocurrency's airdrop or bounty program and complete some simple activities.

Dividends

Cryptocurrency dividends function similarly to stock dividends. You invest in a dividend coin and get a set rate of return. You must purchase and retain cryptocurrencies for the duration of the interest-bearing period.

Crypto Referral Schemes

A crypto referral scheme is a popular and effective way to earn cryptocurrency. Simply register on the website to obtain your unique referral link and begin recommending and earning. You may make a lot of money by promoting DeFi ICO ventures from the comfort of your own home.

Ways Trading Bot Helps You Earn More Money

- It removes emotions from the equation

Ask any seasoned trader about the most critical things for crypto newcomers to bear in mind. We're confident one of them will be avoiding FOMO and FUD.

Controlling your emotions is essential for good trading. However, this is easier said than done. It might be difficult to remain cool and composed when you see your hard-earned money being devoured by paper loss.

Emotional control is a talent that may be honed through time. Meanwhile, you may utilize a cryptocurrency trading bot to assist you in removing your doubts and concerns from the equation. As a consequence, your portfolio will profit from being entirely managed by well-founded, strategic, and non-emotional judgments.

- It uses AI to make a sound investment decision

There is a reason why artificial intelligence is taking over nearly every area of modern life. Massive amounts of data are now available to process, utilize, and fully exploit. Machine learning advancements have drastically enhanced our lives, from how we buy to managing our day-to-day duties.

There's no reason why the same advancements shouldn't be applied to cryptocurrency, and crypto trading bots are meant to accomplish precisely that.

Stoic, for example, rebalances top-20 crypto assets using projections from 156,000 analysts registered with Cindicator and AI algorithms. You cannot accomplish this independently, and your crypto trading decisions will become significantly wiser once you begin utilizing Stoic. You can automate the harvesting of the expertise of thousands of specialists.

- It makes scaling your investments easy

Assume you learn everything you need to know about crypto trading on your own and can devote several hours to it every day. Many things will continue to limit your earning potential.

For one thing, the human body is only capable of so much in a single day. Even if you make cryptocurrency trading your full-time profession, you'll still need to eat, sleep, and perform other everyday tasks. Realistically, the maximum time somebody can devote to crypto trading is 8-12 hours each day.

The beauty of the crypto market, though, is that it is open 24 hours a day, seven days a week. This means that income possibilities might present themselves at any time. With a crypto trading bot like Stoic, you can feel confident that transactions will be performed when needed, boosting your profits even more.

Increasing your investment with a crypto bot is as simple as adding additional money to your account. The bot will automatically decide where to spend the money. Scaling your investments is therefore made easier than ever before.

Takeaway

Artificial intelligence and machine learning have changed the way we live our lives, and they're only going to become better in the coming years. Any astute trader understands that AI and machine learning can provide a plethora of information that may help them optimize their profits – and this is best accomplished using crypto trading bots.

Best Crypto Trading Bot to Earn

Crypto trading bots allow you to manage all of your cryptocurrency exchange accounts in one spot. These apps simplify trade for Ethereum, Litecoin, Bitcoin (BTC), and other cryptocurrencies.

The following is a chosen selection of the top crypto trading bots, along with their most popular features. The software on the list is both open source and commercial.

Pionex Inc

Pionex is a cryptocurrency exchange with built-in trading bots. For no additional cost, you may use 12 different trading bots. These bots enable you to automate your investment approach, removing the need for you to watch the market continually. Pionex has cheap trading commissions and a fully functional mobile app, and Pionex would be an excellent choice for high-volume, mobile investors.

Users

- High-volume investors seeking the lowest costs
- Cryptocurrency investors on the go
- Traders with limited time
- Those seeking a low-risk investing plan

How to Get Started with Pionex Crypto Trading Bot

The most well-known cryptocurrency trading bot for traders is the Pionex Grid.

How Grid Trading Bot earn money

- Traders can profit from Bitcoin investments. When it comes to investing, the outcome is mainly determined by your trading ability and risk tolerance. Grid trading bot assists you in buying low and selling high—nothing else.
- If you have a cryptocurrency trading strategy in mind, this is one of the most excellent tools. You don't have to stare at the chart all day; instead, let the bot assist you in executing trades.
- Crypto trading bots operate in both a turbulent and an up-trending market, and it won't work if the price keeps falling and falling.

Product Selection at Pionex

Even though Pionex offers manual trading via crypto-to-crypto conversions, its primary offering is its trading bot selection. When certain pre-defined market conditions are satisfied, a cryptocurrency trading bot is an automated software that executes buy and sell orders with no operator input.

Pionex presently provides 18 different automated trading bot choices. Let's look at some of the company's most popular products.

- Grid trading bot: The Grid trading bot buys an asset when the price reaches a predetermined "low" point and sells when the price reaches a predefined "high." This function enables the user to profit from market volatility constantly.

- Leveraged grid bot: The leveraged grid bot works the same way as the conventional grid bot, but it allows you to use margin to boost returns. Leverage levels currently accessible include 1.2x, 1.5x, 2x, and 3x.

- Trailing sell bot: The trailing sell bot was created to appeal to investors who have a habit of selling too soon in a rising market. The trailing sale bot allows you to define numerous target prices that will rise in value over time. The bot will sell a predetermined percentage of your assets when each goal is met. This strategy also lowers possible losses if the market abruptly reverses course.

- The DCA (dollar-cost averaging) bot allows you to set a series of buy orders at time-based intervals. Rather than basing purchase order execution on price, the bot will buy at the current market price at each of your intervals. This concept gives a straightforward method for achieving a higher average price over time. Pionex presently supports periods of 10 minutes, 1 hour, 1 day, 1 week, and 1 month.

- Spot futures arbitrage bot: This bot allows you to profit from discrepancies in the current spot price and the current price in the futures market. This bot promises a "return of 15-50 percent APR with extremely minimal risk," and you can get started with just a $50 payment.

- Martingale bot: This bot conducts DCA buy and one-time sales to profit from fluctuations.

- Rebalancing bot: This bot assists you in hoarding money.

Pionex offers manual trading between some of the most prominent cryptocurrencies in addition to trading algorithms. Traders may use the Smart Trade terminal to establish a stop-loss, take-profit, and trailing stop in a single trade.

However, the site only offers crypto-to-crypto trading; unlike other exchanges, you cannot fund your account with fiat cash through a credit card or bank transfer. If you don't already have cryptocurrency, you'll need to register an account with a broker that supports fiat transfers (such as Kraken or Coinbase), purchase a base currency such as Bitcoin or Tether, and then transfer the base currency to your Pionex wallet.

Pionex Pricing

Pionex has a maker-taker fee structure, which means you'll be charged a fee when you perform transactions that "create" liquidity on the market and "take" liquidity away from the market. You must pay the maker charge if an outstanding order on the books does not promptly match your transaction. If you place a trade order immediately matched with an outstanding order, you will be charged the taker cost. Pionex's maker and taker fees are now fixed at 0.05 percent of the entire deal value. Your fee is deducted from the total outstanding sum put into your account when your deal is completed.

If you are a high-volume trader, you might consider applying to Pionex's Market Maker program. To be eligible for the Market Maker program, you must deposit at least 300,000 in USDT into your account. You can also apply if you have the equivalent of 300,000 USD or more in another supported currency or token (Bitcoin, Chainlink, etc.). If you are eligible for the Market Maker program, you will pay no maker costs, and taker fees remain unchanged at 0.05 percent.

The Market Maker program is only available for the first ten days each month. Once approved into the program, you must maintain a total account balance of $300,000 or more on a continuous basis. If you are rejected from the program due to a low balance, you must wait at least 30 days before applying to become a Market Maker member.

Customer Service at Pionex

Pionex has a few distinct means for contacting their support team.

- Email: Send a mail to Pionex official mail address.
- Live chat: To contact Pionex's customer care department, go to the broker's site and select the "Support" button. Then, select "Live chat" from the drop-down menu. A popup window will open on your screen, allowing you to contact a representative.

The hours of operation for Pionex's customer support staff are not presently listed on the company's website. When we attempted to use the live chat capabilities at about 10:30 a.m. EST, we were advised that their customer support team was unavailable. You may, however, find valuable instructions on the broker's YouTube channel. We'd like to see Pionex expand its customer support services to include phone chat in the future, as well as specify its operating hours.

Pionex Mobile App

The Pionex mobile app is completely functional, replicating all of the capabilities found on the desktop platform. We found the user experience easy enough that novice users will have no trouble exploring features, and we were even able to put up our trading bot in under a minute.

In addition to bot functionality and manual trading, you'll be able to use the app to access the following features.

- Simple manual trading: Conduct you like to make your investment by hand? The Pionex app also allows for simple manual order placing.
- An all-encompassing, responsive user interface: It is critical to access up-to-the-minute market data and pricing information in the continually open bitcoin market. We saw no latency between Pionex's desktop and mobile apps, implying that traders may obtain trustworthy information regardless of where they invest.

The Pionex mobile app is free to download for Android and iOS devices. Tablet users can also access a proprietary application.

User

- Mobile traders who conduct the majority of their trading on the go
- Traders who forsake their trading strategy due to emotional reasons
- Traders who already have a form of cryptocurrency.

Pionex Safety & Security

Pionex has more than $10 million in funding from Gaorong Capital, Shunwei Capital, and ZhenFund. Furthermore, it has obtained MSB (Money Services Business) authorization from the Financial Crimes Enforcement Network, a US federal body that monitors online financial transactions to reduce fraud and money laundering. Pionex does not educate consumers about the encryption mechanisms to avoid data breaches, something we would want to see the broker implement in the future.

Features:

For retail investors, Pionex offers 12 free trading bots.

- Compared to most main exchanges, the trading fee is the lowest. The trading cost is 0.05 percent for both the maker and the taker.
- Grid Trading Bot enables users to purchase cheap and sell high within a given price range.
- Grid Bot with Leverage delivers up to 5x leverage.
- The Spot-Futures Arbitrage bot enables ordinary investors to generate passive income with minimal risk. This strategy's anticipated return is 1550 percent APR.
- Smart Transaction terminal enables traders to establish stop-loss, take-profit, and trailing stops in a single trade.
- MSB (Money Services Business) License issued by US FinCEN Price: Free
- Exchanges: Binance, Huobi Global, Pionex

Pros

- 12 trading bots that are free to use
- All transactions have low fees
- Mobile offering that is well-designed

Cons

- Accounts cannot be funded using fiat cash.

Cryptohopper

Cryptohopper is an all-in-one trading platform designed to make investing in the cryptocurrency market simpler and more intuitive. Investors may use Cryptohopper to automate their investment procedures, replicate professional traders, establish signals and alerts, and much more. This enables investors to keep track of the market regardless of their trade currencies, the exchanges they use, or where they travel.

Although investors must pay to utilize Cryptohopper's most advanced features, the site does not impose a per-trade commission. Instead, it charges investors a set monthly fee based on the number of tools they have access to.

A cryptocurrency, unlike the stock market, may be created by anybody. As a result, cryptocurrency trading platforms have the freedom to support any coins they deem suitable. As a result, numerous exchanges and brokerages will offer various cryptocurrencies, forcing investors to open several accounts with different platforms to invest in multiple altcoins. Managing many cryptocurrency accounts may be difficult, and Cryptohopper tackles this problem by combining your accounts into one location. Continue reading to discover about Cryptohopper's distinct features and if it's worthwhile to open an account with the cryptocurrency aggregator.

Cryptohopper is best for Investors interested in copy trading and already have accounts with their preferred brokerages. It is also good for Investors seeking advanced analysis tools that do not necessitate professional coding expertise to implement

Cryptohopper Fees and Account Minimums

Most investors can use Cryptohopper's platform for free. Track your holdings, execute real-time trades through Cryptohopper's trading terminal, and use strategy backtesting features for free. There is no minimum amount required to begin using Cryptohoppe. Still, the platform does recommend depositing at least €300 to your preferred exchange so that Cryptohopper can spread risk across multiple transactions while also meeting minimum trade requirements.

If you want to use Cryptohopper's automated investing features, you must upgrade to a paid account. Currently, Cryptohopper offers three tiers of paid account access.

- The Explorer package comes with 1 simulated trading bot, 80 positions, a maximum of 15 chosen coins, and 2 triggers. For access to the Explorer plan, you may expect to spend around $19 per month.

- The Adventurer package is Cryptohopper's intermediate account tier. The Adventurer package contains all of the features included in the Explorer package, plus access to 50 more coins, 5 maximum triggers, and exchange arbitrage. You'll have to pay $49 per month to get access to the Adventurer package's features.

- If you're a seasoned investor seeking the most comprehensive set of tools available, Cryptohopper's all-inclusive Hero package may be for you. With the Hero package, you may open up to 500 positions simultaneously with up to 75 different currencies, using up to 10 triggers and infinite trading signals. You'll also have access to all of Cryptohopper's premium features and extensive technical analysis capability. The Hero bundle is available for $99 per month.

Remember that Cryptohopper is not an exchange; it is a trading platform that allows access to different marketplaces. This means you'll have to pay the costs levied by the exchange where your order is completed, as well as any monthly fees levied by Cryptohopper.

Ease of Use

Cryptohopper is a tool for automatic bitcoin investment. Though you'll need to subscribe to a subscription plan to use Cryptohopper's main features, you may start exploring the site for free - you don't even need to provide your credit card information until you're sure you want to commit to a plan.

Even though Cryptohopper focuses on providing investors with access to more complex charting tools and trading possibilities, the site remains straightforward to use. Some of our favorite aspects that make Cryptohopper, so simple to use are:

- Expert-level tools with no coding requirements: Unlike most open-source platforms, Cryptohopper allows you to customize your investing experience and gain access to high-level tools without having to code them yourself. Cryptohopper combines strong capabilities with a simple platform for novices, from unique strategy scanners that rapidly spot candlestick charting patterns to backtesting investment ideas.
- Setup is quick and straightforward: To get started with Cryptohopper, you don't need any prior cryptocurrency understanding. Without having an exchange account, you may set up a paper trading account to test out Cryptohopper's capabilities. Most people may establish an account and get started in as little as three minutes.
- There are no extra fees: Cryptohopper is not a broker, and the platform does not charge a commission on your trades. You'll always pay the exact, modest standard cost no matter how many

trades you make every month. This also makes calculating how much your exchange will cost in fees easier because you don't have to account for additional expenses.

Overall, we rated Cryptohopper high scores for its platform's simplicity. With a comprehensive trading terminal and an easy-to-use interface, even complete beginners can start investing quickly with Cryptohopper.

Cryptohopper Education

Cryptohopper not only lets you manage all of your exchange accounts from a single platform but also provides newcomers with a variety of unique materials to help them learn more about investing and the cryptocurrency market in general. Let's take a look at some of our favorite educational tools and materials offered by Cryptohopper.

- Paper trading accounts: With Cryptohopper's paper trading accounts, you may test out your investment plan without risking any of your money. Paper trading access can also be useful for people who are still debating whether Cryptohopper is the perfect trading terminal for them and do not want to commit to paying for an account just yet.
- Professional-grade analysis tools: The variety of indicators and analysis tools provided by Cryptohopper is excellent. Depending on the package plan you select, you will have access to up to 90 candlestick pattern recognitions, 30 distinct indicators, and customization features to customize your automated trading experience. These features don't require any coding knowledge to set up. Click to configure your alerts and indicators.
- Simple copy trading functionality: Cryptohopper's Marketplace gives users all the tools they need to learn from some of the market's most successful cryptocurrency investors. Cryptohopper personally analyzes each candidate to guarantee that you are learning from the best, and you can instantly buy

bot templates, communicate with other investors, or subscribe to trade signals from your phone or desktop application.

Customer Support at Cryptohopper

Cryptohopper provides consumers with several options for contacting customer service:

- Using internet chat: Go to their website to use Cryptohopper's online chat. To access live chat, click the blue speech bubble in the lower right-hand corner of the home screen. You may connect to live chat or view the company's FAQ directly from the chat app. Though you'll be linked to a chatbot at first, you can contact a team member if your answer isn't located in the current FAQ.
- Using the support ticket service: Click here to open a support issue with Cryptohopper. To utilize ticketing customer support, you must already have a Cryptohopper account.

There is currently no mobile option for customer support.

Cryptohopper Offerings

Cryptohopper is an all-in-one trading platform that enables investors to access their cryptocurrency brokerage accounts from a single platform. Cryptohopper offers investors a variety of tools and services to help them improve their on-exchange skills, including:

- Copytrading capabilities
- Access to strategy designer tools and tools that allow investors to profit from pricing disparities across exchanges
- Tools and functionality for advanced analysis
- Bear market tools like shorting tools and dollar-cost averaging

Investors may use these capabilities without fear of losing money on each transaction because Cryptohopper charges a flat monthly price depending on the tools offered at each tier. Though Cryptohopper does not provide brokerage services, it does supply investors with a full collection of tools that may be used to make trading more straightforward and more intuitive.

Cryptohopper Mobile App

Cryptohopper's quick setup and all-inclusive access are now available on Apple and Android mobile platforms. Some of the things we like about the Cryptohopper app are as follows:

High-level security features include:

- Biometric security mechanisms are built into the Cryptohopper app to ensure that only you have access to your assets.
- Optional push notifications: Configure push notifications to get real-time updates on specified triggers, patterns, outstanding orders, and other events. You can track your progress and positions from your phone's home screen.
- Orders with a single click. Cryptohopper's mobile platform is equally easy to use as its desktop counterparts. Execute trades using a single order and manage your positions from anywhere.

The Cryptohopper app is free to download for both Apple and Android devices.

Features:

- You may design your technical analysis with this bitcoin trading robot.
- It is simple to obtain tactics and bot templates.
- Cryptohopper safeguards your account with security mechanisms.
- This free bitcoin trading bot protects your personal information.
- It has an easy-to-use UI.
- The program generates performance reports in real-time.
- Apps for iOS and Android

- Algorithms support EMA, RSI, BB, and other typical signals/indicators.
- Free of charge

Some of the exchanges include HitBTC, Okex, KuCoin, Bitvavo, Bitpanda Pro, Huobi, Poloniex, Kraken, Bittrex, Bitfinex, Coinbase Pro, Binance.

Pros

- Does not impose a commission on monthly payments.
- Access to high-level analytical tools for investors
- Advanced security protections are included in the comprehensive mobile app.

Cons

- There are no phone customer service alternatives.
- You must have a monthly subscription to access the most beneficial features.

Bitsgap

Bitsgap is one of the top bitcoin trading bots that allows you to manage your crypto holdings easily. This tool can evaluate over 10,000 cryptocurrency pairs and choose the coin with the most potential. It lets you build your bot strategy with only a few mouse clicks.

Bitsgap is an all-in-one cryptocurrency trading platform that allows you to trade, manage your portfolio, use trading bots, get alerts, and much more. It has also integrated over 25 major exchanges, including Coinbase Pro, Binance, Poloniex, Kraken, Bitfinex, etc.

The sign-up procedure will take you two minutes because it only takes an email account verification. Users can also join using their Google or Facebook accounts. Furthermore, Bitsgap offers a 14-day free trial during which the trader may take advantage of numerous services for free.

Features

The following features are simply the tip of the iceberg for all of the amazing, ready-to-use capabilities available on the platform.

- Arbitrage is the practice of maximizing the price difference between two or more transactions.
- Trading Terminal– High-quality tools are provided to assist you in trading on several exchanges from a single interface.
- Signals – keep an eye out for any market irregularities to make the best trading judgments.
- Portfolio Tracking– Traders will have access to a live-generated portfolio that will allow them to keep track of their funds and earnings.
- Automated trading bots — Bitsgap offers pre-configured trading bots optimized using machine learning.
- High Security - Customers connect their exchange accounts to Bitsgap using the exchanges' API credentials; no deposits are made on the site itself. Bitsgap never has access to the customer's cash; it simply executes trades on the user's behalf. Bitsgap secures all API orders with high-end 2048-bit encryption for additional security.
- Demo Exchange– 'My exchanges' displays all exchange accounts the user is connected through API. The 'Demo' option, on the other hand, allows the trader to practice utilizing all of Bitsgap's distinctive features, allowing the user to become acquainted with all the platform has to offer.

How Bitsgap Works

Let's have a look at how Bitsgap works:

- Create an account on Bitsgap first.
- Using your API credentials, connect your Bitsgap account to one available exchange.

- Simply make an order, launch a bot, or take advantage of arbitrage chances on the chosen exchange using Bitsgap.

Trading Terminal

If you've ever traded on a cryptocurrency exchange, you're probably aware that exchanges prioritize liquidity above giving the greatest trading tools.

As a result, systems like Bitsgap focus on giving traders the greatest crypto experience possible. The Bitsgap trading terminal offers a unified interface for trading on several cryptocurrency exchanges.

Trading features such as take profit, shadow orders, limit market, and stop-loss provide you a competitive advantage when making market orders.

Bitsgap Grid Trading Bot

The Bitsgap trading bot is driven by the grid algorithm, which lets the trader establish investment ranges and limitations. Then it proceeds to disperse these investments proportionately. As a result, whenever the system executes a limit order, the grid will put another sell order at a little higher price than the market value.

Limit orders work after the preceding order is completed. Bitsgap proceeds to make another purchase limit slightly lower than the market value. As a result, as long as the crypto prices do not surpass the trader's preset value range, the system will continue to execute transactions indefinitely.

Bitsgap offers trading bot strategies that have been pre-configured.

- Backtesting

Bitsgap's backtesting function actively optimizes pre-configured trading bots for bull (uptrend) and bear (falling) markets. The team employs machine learning to enhance revenue and avoid risks by examining and analyzing past data. Traders can select any trading pair based on their historical performance and trading bot results.

- Bitsgap Arbitrage

Bitsgap monitors cryptocurrency prices across different exchanges and identifies arbitrage possibilities. Traders might profit from these changes by engaging in crypto arbitrage. Bitsgap allows you to search for different cryptocurrencies and allocate balance when arbitrage.

Bistgap Signals

Bitsgap also displays market signals. These are market price anomalies, which indicate a rapid shift in the price of a certain cryptocurrency.

This enables traders to spot and profit from price swings as they occur. You may also search for these specific crypto assets. Price movement notifications, on the other hand, cannot be configured.

Portfolio Monitoring

Bitsgap makes it simple to keep track of your cryptocurrency assets. You can see your whole portfolio, as well as your ROI (return on investment) and positions. You may also view your open orders and order history. If you use more than one cryptocurrency exchange, you may filter your portfolio for any individual exchange.

Pricing

Bitsgap offers three subscription options. All options include a 14-day free trial. We recommend beginning with the Advance plan if you are a newbie because it contains the arbitrage function. If you can successfully perform an arbitrage, you might gain more than what you paid for the Bitsgap.

Demo Account

Beginners may trade with a Bitsgap demo account with no financial risk. The demo account comes with boosted paper money to facilitate trading. Furthermore, using Bitsgap, you may enjoy the following advantages:

With a Bitsgap demo account, you can simply access the top five cryptocurrency exchanges.

- You can practice trading using the virtual fund's 5 BTC.
- You may select the preferred approach and try it.
- You may obtain trading experience in real-time market circumstances.

Bitsgap Affiliate Program

Bitsgap allows you to earn a 30% commission on every transaction by simply encouraging your friends to join Bitsgap. All you have to do is follow the steps below:

- Copy your unique affiliate link from Bitsgap and share it on social media.
- The more people you can invite; the more money you can earn.
- Earn 30% commission on every Bitsgap recommendation you make.

Bitsgap Security

Bitsgap provides excellent security and protects your assets in your hands. You may use 2FA to increase the security of your personal finances. Furthermore, your monies and assets are only secured and linked through your completely encrypted API keys. Moreover, these APIs enable Bitsgap to conduct transactions and collect data on your behalf.

In Crux, we can argue that Bitsgap is an all-arounder platform with automated trading algorithms that create a consistent stream of revenue regardless of market conditions. Not to mention the other features like stop-loss, take-profit, exit strategies, and following up that secure your safety and greater ROI (Return-on-investment). Furthermore, its security is excellent, and it offers good customer service.

Binance, OKEX, Bitfinex, HitBTC, Bittrex, Huobi, Exmo, Kucoin, CEX.IO, Kraken, Poloniex, Livecoin, Coinbene, Coinex, Gate.io, Bitstamp, Liquid, Gemini, Bit-Z, Yobit, DDEX, Bithump, Bibox, BigOne are some of the exchanges.

Pros

- Trading bots that are already configured
- Dashboard for Market Signals
- Configuration is straightforward.
- Simple and quick start
- Everything is in one spot.
- Accounts with a limited budget can take advantage of a free package.
- Gives a two-week trial period with no credit card required.

Cons

- There is no mobile app
- Trading bot functionality is limited.

Coinrule

Coinrule allows you to design personalized trading rules, access 150 templates, and take advantage of many low-cost trading plans. If you are a beginner crypto trader, you may test your techniques using paper trading and backtesting. Coinrule is an automated trading software that allows you to trade on Binance, Kraken, Coinbase Pro, and others.

Starter, Hobbyist, Trader, and Pro programs are available for traders of various skill levels. Each plan assists you in developing your abilities and knowledge to become a better bitcoin trader. The platform is one of the finest for learning to be a consistently successful crypto trader and trading digital currencies.

Let's be honest: keeping track of bitcoin investments may be difficult. Unlike the stock market, cryptocurrencies trade 24 hours a day, seven days a week. Because cryptocurrencies may only be listed on individual exchanges, many investors must use several trading platforms. Fortunately, programs have been developed to facilitate bitcoin investment management. You can link all of your trading accounts, from Coinbase to Binance, in one location by using an investing platform like Coinrule. This can help to simplify portfolio management and make it easier to exchange digital assets.

Coinrule is a platform that allows you to trade cryptocurrencies in a secure environment. The platform establishes trading regulations, employs military-grade encryption, and collaborates with the biggest cryptocurrency exchanges.

Coinrule's Product Offerings

The platform allows you to design your trading tactics, which is very useful for high-frequency traders. No coding skills are required to construct simple tools for making customizable transactions using Coinrule. Because Coinrule is not an exchange, you must choose your preferred exchange, and you have the option of backtesting your trading technique.

These rules, usually based on technical indications, may trade bitcoin straight from the Coinrule site. Rules may be changed, and your experiences allow you to devise new techniques. You may put your new trading rules to the test and track their effectiveness to increase your profits.

Coinrule provides up to 150 templates with rule customization based on customizable trading techniques. The program assists you in establishing the method you wish to test and grow on as you create your system. You may test the techniques on the free trial to see how they function and which ones are best for you. The program provides insight into bitcoin trading, which promotes your confidence, understanding, and, eventually, profitability.

Coinrule Pricing

Coinrule has a tiered pricing system for various traders, including the Starter, Hobbyist, Trader, and Pro plans. The Starter plan is completely free. You get two demos/rules, one exchange connection, seven template strategies, unlimited paper trading, and the ability to trade up to $3,000 in cryptocurrency every month.

The monthly fee for the Hobbyist plan is $29.99. You may employ up to 40 template strategies, 7 demos/rules, connections to two exchanges, and access to the trader community. A monthly trading volume of up to $100,000 is authorized.

The Trader package costs $59.99 per month. It includes limitless templates, 15 demos/rules, one-on-one training sessions, access to three exchanges, and the opportunity to trade digital currencies worth up to $1 million. For $449.99 per month, the Pro plan includes 50 demos/rules, unlimited templates/exchanges, a dedicated server, and $5 million in transactions. Each strategy will assist you in becoming a better trader and transitioning into different trading techniques as your abilities develop.

Coinrule Promotions

Coinrule's promotions begin with a free account, which may be upgraded to a subscription plan. When you buy an annual membership instead of a month-to-month plan, you save 20%. Those who sign up for the newsletter enjoy special discounts ranging from 25% to 50% off yearly memberships.

Coinrule Customer Service

Coinrule's customer care reacts rapidly to concerns via chat, email, Facebook, and Twitter. Coinrule employs a blog, FAQs, and a knowledge area to assist clients in locating information without contacting customer support. Customer service responds to queries and collaborates with you to fix any issues. Coinrule contacts users on the same day to inform them of important concerns. The response times are outstanding, and Coinrule's proactive customer support team assists you in dealing with difficulties as soon as possible.

Coinrule Mobile App

Coinrule can only be accessible via the internet, and while the website may be viewed via a mobile device, no mobile trading app is available. The organization aims to expand mobile capabilities to improve the client experience in the future. Users will benefit from a mobile app that will allow them to utilize Coinrule on the go.

Coinrule UX

Coinrule users say it helps them get better at trading cryptocurrency. You may test and fine-tune your technique to find the best purchase and sell points. You may tweak the rules to comprehend cryptocurrency developments better and take advantage of them.

On the left side of the login page, you can see the exchanges, rules, and settings. The exchanges show the platforms and balances, and the regulations outline all of the criteria and techniques for effectively limiting them. All of your options are displayed under Settings. The flexibility to tailor the program to your specific requirements distinguishes Coinrule from its competitors.

Coinrule Security

To secure your personal and trade information, Coinrule employs a robust security infrastructure. This is accomplished by authentication methods, security measures such as denial of service attacks prevention, and payment protection. Coinrule uses algorithms and military-grade encryption technologies to store and confirm essential information.

Coinrule prioritizes the prevention of denial of service attacks. The firm collaborates with CloudFlare to monitor and prevent the platform from going down.

Coinrule does not directly collect payments or personal information; instead, all payments are routed through Stripe. Stripe classifies the payments as merchant-initiated and eliminates the requirement for credit cards. Coinrule can safeguard your data thanks to the additional layer. All of these elements combine to make Coinrule a secure platform. These security elements combine to produce layers that keep hackers from breaching into Coinrule and causing disruptions. Coinrule is a safe platform for cryptocurrency trading and education.

Features:

- Provides customer service in real-time.
- It enables you to test rule performance on historical data.
- Create your own trading rules.
- It contains a market indication that allows you to deploy funds easily.
- Military-grade encryption is provided.
- You can swap at any time.
- Free of charge
- Binance, HitBTC, Coinbase Pro, Okex, Bitstamp, Bittrex, Poloniex, Kraken, BitMEX, Bitfinex, Bitpanda pro, Liquid, Binance US are some of the coinrule exchanges.

Pros

- Excellent client service
- High-level security
- Several price options are available, each with its own set of promos.
- Indicators of advanced trading

Cons

- There is no mobile trading app.

Trality

Trality is a platform for anyone who wants to earn from algorithmic cryptocurrency trading without quitting their day job. They provide cutting-edge tools for developing incredibly complex, highly creative algorithms in an instructional, community-driven architecture that fosters learning and growth.

Trality is a trading bot generation platform based on the Python application programming interface (API) appropriate for novice and experienced cryptocurrency traders.

The Trality web app is completely cloud-based, allowing you to construct without the need for any specific equipment or downloads. Trality's Code Editor tool will appeal to more experienced traders. At the same time, newbies will benefit from its Rule Builder, which allows you to simply develop trading bots by dragging and dropping indicators to design your strategy. With a fresh viewpoint on money and wealth, major cryptocurrencies have transformed the world of finance. Crypto trading bots automate the trading process by generating buy and sell recommendations using algorithms based on technical indicators. When a bot is linked to a trading account, it may automatically start and liquidate deals on cryptocurrency exchanges.

Trality is a firm based in Austria launched in 2018 by Christopher Helf and Moritz Ptzhammer. It has created a platform to construct and configure your cryptocurrency trading bot. The cloud-based method allows you to create a trading bot using your web browser, eliminating the need to purchase additional hardware or download software to your desktop.

Because of its versatility and simplicity, the Trality platform has grown in popularity. In addition to its user-friendly design, the platform encourages crypto traders to study and progress within its community-driven infrastructure.

Another significant feature of the Trality platform is its ability to perform paper trading and backtesting. The Trality platform lets you explore how your trading strategies and algorithms might have fared in the past and then practice trading with virtual money before committing any dollars. This function also allows you to store your backtesting history, as the platform contains many past data. You may also apply a variety of computational testing situations, such as specific time frames.

The Rule Builder and the Code Editor are the platform's two primary tools. The Rule Builder allows you to create a trading bot by dragging and dropping multiple indicators and methods. The Code Editor is designed for traders who are comfortable with Python coding and can utilize their coding skills to create unique algorithms for more complex strategies.

Customer Support

Trality works hard to ensure that all users receive the assistance they require. They provide a Helpdesk where you can search for answers to issues you may have while constructing your bot, as well as a very helpful, community-driven Discord room where its members may ask questions and aid others on their bot-building adventure. Furthermore, if you need to speak with Trality one-on-one, they provide support email addresses for urgent and more personal inquiries at support@trality.com.

Trality Mobile App

Trality does not yet have a smartphone app, although one is reportedly in the works. The business is also working on the Trality Bot Marketplace, which will allow you to either publish your bot or follow bots written by other developers.

User Experience

This trading algorithm creation platform can be useful for Python developers and individuals who don't want to code to create a functioning bitcoin trading algorithm. You may simply design a trading bot with all the functionality you need in a drag-and-drop, graphic-based, and interactive user interface by utilizing the platform's Rule Builder.

The Rule Builder allows you to easily determine how technical indicators trigger trades and where to place your limit and stop-loss orders without coding the strategy yourself. Traders who are experienced with coding can utilize the Python API to modify the code of the algorithms and construct customized bots to apply bespoke trading strategies.

Benefits of Trality

Trality's cloud-based platform allows you to execute trading strategies straight from your Web browser, eliminating the need for lengthy program downloads. Even if you have no coding skills, you can construct trading bots.

Trality also allows you to backtest and test your bot in actual market circumstances using virtual money, eliminating the need to risk your assets until your bot has been completely verified. Trality's Python Augmented Code Editor, which offers a user-friendly UI and good Python API capabilities, is ideal for advanced users. You may sign up with Trality and use virtual money to paper trade for free. The program performs your bot's orders on a bitcoin exchange that supports it.

Trality Security

Because your money remains deposited at your crypto exchange, Trality exposes them to no risks other than the major market risks associated with cryptocurrency trading and the potential that your algorithm's code contains a mistake.

Trality never has direct access to your cash, and the site exclusively uses approved exchange APIs. Your trading bots and strategies are end-to-end encrypted for extra protection, so even Trality cannot access them.

Features:

- Curated, pre-defined strategies
- Graphical UI with drag-and-drop functionality (for beginners)
- Python code editor in the browser (for advanced traders who know Python)
- Backtesting module with lightning-fast performance
- Cloud-based real-time trading. Your algorithms operate consistently and never miss a deal.
- Binance, Kraken, Bitpanda, and Coinbase pro

Pro

- Tools for both programmers and non-coders
- End-to-end encryption
- The system is entirely cloud-based
- User interface that is simple to use, interactive, and graphically based

Cons

- There is currently no mobile application.
- It has the most popular trades, but more are needed.

3commas

3Commas is one of the greatest cryptocurrency trading bots, allowing you to boost earnings while decreasing losses and dangers. This program enables you to benefit with little effort, and it allows you to build a strategy based on 20 trading indicators.

3commas Potential User

- All skill levels

- Those who wish to have complete access to all exchanges
- Investors seeking individualization

3commas Offerings

3Commas has four service tiers: free, starter, advanced, and pro. The following features are included in the free plan:

- Buying and selling on 23 cryptocurrency exchanges by hand
- Viewing and copying trade bots
- Brief algorithms
- Prolonged algorithms
- API access for automated portfolio balancing

You may sign up for the free plan right now and use all of the training, education, paper trading, and other learning tools they provide without having to pay a monthly charge. This is a terrific method to begin started since you'll be able to see if you'll profit from the platform before paying for it.

After completing the courses and developing a trading strategy that aligns with your investment objectives, you can go to one of the more advanced programs. You may start putting your new trade talents to use with these more complex strategies!

The more sophisticated ideas include capabilities like smart sell, options bots, GRID bots, and DCA bots. The ideal plan is determined by how much you intend to trade and if you want to trade options and futures. While the free plan contains some basic functions, you have restricted access to most 3Commas' services. We propose a paid plan if you want to perform more serious trading.

Features:

- In a single window, you may sell and buy coins.
- This platform allows you to trade 24 hours a day, seven days a week.

- Other bots' settings can be copied.
- It helps you to balance your portfolio by keeping coin ratios constant.
- This bitcoin margin trading platform provides deal notifications via a mobile app, internet, and email.
- Backtesting, Dollar-cost averaging, Signals, Custom TradingView Signals, and other features are supported via iOS and Android mobile apps.
- Apps for iOS and Android

Some the 3commas exchanges are Binance DEX, Binance.us, Bilance Futures, Bitfinex, BitMEX, Bittrex, Bybit, Bitstamp, Coinbase Pro, Gate.io, Cex.io, Exam, Huobi Global, KuCoin, HitBTC, Kraken, Poloniex, OKEx, YoBit

Pros

- Extensive teaching and learning resources
- Plans that are both affordable and scalable for traders of all skill levels
- A user-friendly dashboard that contains all of the functions you require in one location.
- Access all major cryptocurrency exchanges from one platform.
- Responsive customer service
- View and copy a library of bots.

Cons

- For newcomers, the onboarding process is lengthy and somewhat overwhelming.
- There may be some little latency in the mobile app.

3Commas Trading Terminal

If you are a newbie, we urge you to begin with the fundamentals by understanding the 3Commas trading terminal interface. This is a brand new section in the user's account that will teach you the fundamentals of trading. The features of the 3commas trading terminal are the focus of this section, and it will assist you in navigating the interface effectively.

Features of the Interface

Let us look at the major features of the trading terminal interface.

- The terminal for placing orders

- The order book (market depth)

- Chart of the traded asset with Tradingview features

- Account Portfolio, Open Orders, and Trade History menu.

Terminal

Here, you specify the parameters of your order (purchase or sale of a cryptocurrency) to be placed on the exchange. Trade is performed when it corresponds with the opposing order at the stated price.

- The Type of order (Market or Limit order). The former is performed at the best price possible on the exchange.

- Stop Loss button. You may quickly establish a Stop Loss order with this button to cancel the trade at a Trigger price if the asset's price swings in the wrong direction.

- Units and Totals fields. This is the o Order size in cryptocurrencies from the trading pair.

- You may use the Conditional switch to add extra parameters to both market and limit orders. You may set the Trigger Price level in the first scenario, and when it is achieved, the market order is instantly issued for execution.

- Order size buttons. With this button, you may rapidly specify the order size (25 percent, 50 percent, 75 percent, or 100 percent of available cash) using these.

Now let us move on to the next part.

Order Book

Market depth is a collection of traders' actual requests (orders) to buy and sell an asset. The order book is commonly shown in different exchanges; the 3Commas trading platform displays it as a continually updating table reflecting market depth.

Let us take a deeper look at the order book's components:

- Management of the displayed orders list. When the first button is pressed, both sell and buy orders are shown; when the second button is pressed, only buy orders are shown; and when the third button is pressed, just the sell orders are shown.

- Sell orders list. Here, the offers are displayed in increasing order from the lowest possible price to the highest. The offers at the bottom of the list are the most advantageous for the buyer. Each line contains both coins' order completion market price and order size.

- Buy orders list. The most lucrative price for the seller is at the top of the list, which is generated similarly.

Trading Chart

The Tradingview platform is used to power the 3commas trading terminal, which shows a specified trading pair chart. It provides a plethora of charting possibilities and the ability to save, alter, and use different technical analysis tools.

The following options can be found in the trading chart section's top menu:

- Timeframe. This value sets the time gap between each candle on the chart. If the time is set to 1 hour, each candle will indicate a 1-hour price change.

- Candle. Heikin Ashi, Bars, Candles, Hollow Candles, Line, Area, or Baseline The chart is presented in Candles by default.

- The Indicators option helps you add indicators to the chart.

- The Undo and Redo buttons help you undo or redo previous activities.

- The Save button saves the chart to the cloud.

- Chart display settings.

- Full-screen mode.

Trades History

One menu contains the Account Portfolio, Open Orders, and Trades History. The tabs at the top allow you to flip between them. The Account Portfolio option shows the cryptocurrencies in your account, their amount, the value change over the past 24 hours, and the market price. All of the orders you have made so far are displayed in the Open Orders menu. The components of the trade history are listed below.

- Trading Pair

- Side (Buy or Sell)

- The Order size and its execution

- Order Type

- Execution Price

- Trigger Price (the price that will cause the order to be executed);

- The timeout option

- Date of order placement

- Buttons to cancel, execute immediately, or renew an order.

- The Refresh button in the menu's bottom-left corner helps you refresh the list of orders. At the bottom-right of the terminal, some page display options enable you to select the number of shown orders per page.

- The last tab enables you to view the trading history.

Placing Your Order

Now is the time to put theory into practice! We highly advise you to change to Paper Account in the top menu of the website if you are just starting to engage with the trading terminal. In the account settings, you may restore the volume of virtual cash.

Let us start by utilizing the Basic option to place a market order. Press the Buy button after selecting the required quantity of Units. Once the order is completed, it will be shown in your transaction history.

It is worth noting that if you execute fairly significant market orders, you will need enough liquidity. Suppose the most lucrative of the possible limit sell orders from the order book fails to match your market buy order entirely. In that case, the order will automatically progress and be executed through the next limit sell order, and so on, until it is fully covered. This is referred to as slippage. A market order may not always close at the precise price reported. When the execution speed is essential than the price, you should place such orders.

The market order in Pro mode gives you additional options and enables you to fine-tune your trading approach. When the Conditional option is turned on, the Trailing and Timeout options are activated. If the asset's market price surpasses the Trigger price level, the first option enables you to open an order at a more advantageous price stated in percent or the value itself. For example, let us say you set Trailing to 1% of a Market purchase order. The asset's price falls by 7%, passing your Trigger price. The order is set to be made, and the Trailing feature will keep the price falling ill it reverses by 1%, at which point the order will be completed on the exchange account. In this situation, the Timeout option will provide a time interval in seconds over which the activities indicated above will not be performed. In addition, for clarity, the Trigger price is shown as a line on the chart.

Another example is the Basic mode's limit order. Everything is as straightforward as it gets: you select the chosen order completion price, the size, and the Stop Loss. You will be transferred to the Market Order after hitting the corresponding button. You place a limit order first, and then a market order with a Trigger Price is used to execute the Stop Loss. The Stop Loss order will be implemented immediately after the asset's price passes this line.

In Pro mode, the last example is a limit order. In this situation, we may select the Trigger price, after which the order is placed in the Order Book after reaching the stated Order price. The outcomes of our activities may be seen in the Trades History.

All being said, before you begin trading with real money, we suggest that you familiarize yourself with the trading terminal by utilizing the Paper Account option. The 3commas trading terminal is currently in beta, and additional features will be added soon.

Quadency

Quadency is a digital asset management tool that offers institutional and individual traders an automated trading solution. It simplifies the procedure for investing in cryptocurrency.

Bots and Techniques

In terms of functionality, Quadency is jam-packed with choices for creating a customized and accurate trading strategy. The offering is broad enough to provide low-risk general methods appropriate for newbies and specialist bots that seasoned industry professionals may modify to their own needs and leverage particular market patterns.

MACD, Mean Reversion, Bollinger Bands, Accumulator, Grid Trader, Portfolio Rebalancer, Market Maker (+), and Smart Order are Quadency bots. It is instantly clear that Quadency went to great pains to equip traders with an enormous list of tactics. Furthermore, there is a high level of customization available for individuals who want to modify and fine-tune various parameters.

In other words, for any historically recorded pattern in the crypto market during the last decade, there is a suitable technique in Quadency's library that can convert it into profits with the aid of Quadency's bots. Extensive evaluations performed by Quadency's algorithms frequently show the most successful tactics on their own. If that isn't enough, there are signals sourced both within and outside Quadency's community, so you may monitor top-performing strategies on the fly. Quadency's toolbox has something for everyone, which is positive.

Features:

- This program allows you to modify bots easily.
- Offers many automated tools and powerful TradingView (social network) charting.
- A high degree of customer service is available.
- You may use it to trade on Bittrex, Binance, and Kucoin.
- It provides a wide range of trading bots that can be easily modified.
- Allows for automated trading.
- Free of charge
- Apps for iOS and Android are in the works.

Exchanges include AAX, Binance, Bittrex, Kraken, KuCoin, Liquid, OKEx, Binance US, Bitfinex, Coinbase Pro, Gemini, HitBTC, and Poloniex.

Pros

- An all-in-one cryptocurrency trading platform.

- Trading bots that are automated.
- Manager of visual portfolios.

Cons

- There is no mobile trading app.

CryptoHero

CryptoHero is a free AI-powered crypto trading bot that is easy to set up and use for cryptocurrency newcomers. It can enable you to automate your trades effortlessly and for free directly from your phone. No coding knowledge is required.

Users

- Traders who are short on time or do not trade professionally
- Travelers who don't have a lot of time to investigate their investments
- Traders interested in learning more about cryptocurrencies

Cryptohero Pricing

CryptoHero costs extremely little for the vast array of choices available when utilizing their crypto trading bots. You pay nothing for the Basic service, and the Premium service costs $13.99 per month or $139.99 per year. Along with the additional features offered by the Basic plan, the Premium plan includes 15 bots and 3 API connections.

A redesigned referral rewards system is in the works, which will allow you to suggest friends or colleagues and receive incentives that may be reinvested in the platform.

Features:

- Support for multiple exchanges. Connect your APIs with ease.
- An easy-to-use interface
- Track your cryptocurrency trading performance individually or

aggregated across many exchanges using a single app.

- Preset Technical Indicators
- Most popular bots are easily set up with only one click with their discovery tool.
- Before launching, perform a quick backtest using past data.
- With a single click on CryptoHero's Trading Terminal, you can easily access all of your exchanges.
- Users may get technical help 24 hours a day, seven days a week via live chat.
- Web, iOS, and Android versions are all available.
- Apps for iOS and Android

Binance, Huobi, OKEx, Bittrex, Coinbase Pro, FTX, Kraken, Kucoin, Bitfinex, Gate.IO are examples of Cryptohero exchanges.

Pros

- Subscriptions, both basic and premium
- There are several active bots.
- AI enhancement
- Several trading techniques
- Trading terminal with leverage and spot options
- All platforms are supported, including iOS, Android, and the web.
- You may pay with cryptocurrency.

Cons

- Subscriptions for the web and mobile are charged differently.
- The bot home page appears to be simple.

Shrimpy

Shrimpy is a cryptocurrency-specific social trading platform. This program includes automated trading algorithms that can help you boost performance while lowering risk. It assists you in developing a portfolio plan, tracking performance, and monitoring the market.

Users

- Novice crypto investors

Shrimpy.io offers a variety of pricing options. The Starter plan costs $19 per month for spot trading and connects to 5 exchanges while automating 3 accounts and using a 15-minute refresh. The professional package costs $79 per month and includes 5-minute refresh choices, connection to 10 exchanges, automation of 5 accounts, and free optimization and rebalancing. This strategy will eventually include futures trading.

The enterprise subscription costs $299 per month and includes 1-minute balance updating, 10-account automation, connectivity to 25 exchanges, priority support, and a server. These plans are affordable and provide a variety of alternatives. The main disadvantage is that individuals who want extra services and access, such as connecting to more than 25 exchanges, must pay a premium.

Advantages of Shrimpy to a User

Shrimpy simplifies your life by connecting your cryptocurrency accounts to several exchanges and integrating them with social trading. The platform enables you to increase your trading efficiency and keep track of what is going on with your accounts. Using the social trading feature, you may mimic the greatest traders. No other platform is as complete or provides all the tools and features needed to simplify cryptocurrency trading.

Features:

- FIPS (Federal Information Processing Standards) 140-2 is used to encrypt and store API keys securely.
- It has the potential to make portfolio management easier.
- Allows you to see the current market price.

- Provides a dashboard with details about each asset and performance measures for a portfolio.
- KuCoin, Binance, Binance US, Bittrex, Bittrex Global, Coinbase Pro, Kraken, Poloniex, Gemini, Bitbox, BitMart, Huobi Global, HitBTC, OKEx, Bitstamp, Bitfinex are some of the shrimpy exchanges.

Pros

- Shrimpy has a slew of one-of-a-kind features centered on portfolio management.
- The platform is simple to use and comprehend.
- Profitable traders may be copied, and their expertise can be leveraged.

Cons

- Shrimpy does not provide free plans for newcomers.
- Live customer service is not accessible in person.

TradeSanta

TradeSanta is one of the greatest crypto trading bots, allowing you to control your risk easily. This program lets you select the approach that best matches your trading style, and it enables you to establish your profit objective and close the trade at the appropriate time.

You may utilize automated investing tools to trade bitcoin around the clock using TradeSanta's cryptocurrency trading software, eliminating the need to regard trading as a full-time job. The cloud-based technology connects to your exchange and puts orders on your behalf, depending on the unique parameters you provide. This makes automatic trading as simple as a few clicks, and you can check in or make adjustments using the mobile app from anywhere you are, anytime you choose.

Best Beginner crypto or automated investors that want to get started quickly and learn as they go. Intermediate to experienced traders who understand crypto markets and create strategies but just want a tool to execute bot-programming.

Tradesanta Pricing

TradeSanta has three subscription packages available at three distinct price points. The only variation between the bundles is the number of bots you receive. The platform's additional features are available to members at all levels. So, all you have to do is determine how many trading pairs you wish to trade (because each bot is assigned to a specific trading pair).

The following is the cost for each plan:

- $14 per month for the basic package
- The monthly fee for the Advanced Package is $20.
- The most expensive package is $30 per month.

You may join up for a 3-day free trial. While this will let you get inside and look at the dashboard and features, it will not provide you with enough time to develop a feel for the platform. Because the trial period is limited, sign up for it in three days if you know you'll have additional free time to study the site and truly dive in, so you'll know whether you want to commit to a premium subscription.

Customer Support

Overall, TradeSanta customers are pleased with the service they receive. TradeSanta offers live chat help in-platform, typically quick to react. You may also direct your inquiries to the TradeSanta Telegram channel, which is quite active. Finally, you can contact customer service through their official mail.

Tradesanta Mobile App

The mobile app has the same tools and capabilities as the desktop dashboard but a more mobile-friendly layout. Each active bot has its own "bot card," where you can monitor its performance and make changes to its parameters and indicators.

You can also keep track of your whole bitcoin portfolio, build new bots, adjust your subscription plan, and do other things. TradeSanta's mobile app is exactly as straightforward to use as the desktop version. As a result, you may select your choice or easily move between the two.

Tradesanta Security

While bitcoin trading is risky, the TradeSanta platform is relatively safe. It employs two-factor authentication for login, and the API key you provide into your exchange account allows just trade permissions, not withdrawal capabilities. In other words, TradeSanta bots can make orders for you but cannot remove cash from your exchange account.

Features:

- You may quickly deploy trading bots with a short and long strategy.
- Support is available 24 hours a day, seven days a week.
- It enables you to buy or sell a huge volume of cryptocurrency with ease.
- Provides a large number of pre-set templates.
- Bot can be tracked in real-time.
- Price: $14 per month for the Basic Package
- Monthly fee for the Advanced Package is $20.
- The most expensive package is $30 per month.
- Apps for iOS and Android
- HitBTC, Binance, Bitfinex, Bittrex, Huobi, BitMEX, OKEx, and UPbit are tradesenta exchanges.

Pros

- A streamlined method for creating a bot with your indicators and parameters.

- A clean, basic interface makes it simple to track performance, make modifications, and keep track of your portfolio.
- Subscription plans with tiers that are reasonably priced

Cons

- The free trial period is only three days long.

Zignaly

Zignaly is a trading terminal that may be used 24 hours a day, seven days a week. It includes mining Hamster (a cryptocurrency market monitoring service) and crypto quality alerts for trading.

Features:

- This program is compatible with any installation.
- It allows you to divide your earnings objectives.
- You may view your results based on your position.
- Your coins are stored on the exchange by Zignaly.
- Free of charge
- Zignaly exchanges include Binance, KuCoin, BitMEX, FTX, and VCC.

Pros

- A cloud-based solution enables users to use bot functions 24 hours a day, seven days a week, from any location, even if their primary device is not available.
- Bot includes actual analytics such as monthly sales, burn, and churn rates, usually lacking in other trading bots.
- The platform offers simple diagrams with a simplified display of all measurements, statistics, and specifications.
- The Zignaly platform's supported exchanges have been upgraded, and it now includes four crypto exchanges: Binance, KuCoin, VCC Exchange, and BitMex.
- Zignaly trading bot offers services with a $0 beta plan every

month. There are no transaction costs, and however, there are exchange fees.

Cons

- The Zignaly platform only supports a few exchanges. Only four exchanges are available, with the most popular being the well-known Binance exchange.
- Because Binance intends to quit servicing US cryptocurrency traders, Zignaly services may only become limited to non-US consumers.
- Even though Zignaly has free access to the signal portfolio, customers must still purchase signals from the signal suppliers.

Botcrypto

Botcrypto is a platform that allows you to trade cryptocurrency automatically. It enables you to program and operate trading robots to carry out strategies.

Features:

- It allows you to save time.
- You do not need to know how to code to utilize this platform
- Instant backtest allows you to validate your trading strategy.
- A graphic editor may be used to construct your approach.
- Give help 24 hours a day, seven days a week.
- Free of charge
- Exchanges include: Binance and Kraken

Pros

- Simple to use interface
- There is no installation required.
- Backtests and simulations are completely free.
- Strong community
- Learn to trade courses

- Using a visual editor, you can create strategies.

Cons

- Only a few swaps are permitted.
- The Botcrypto shop has a limited number of strategies.

Kryll.io

Kryll.io is one of the top cryptocurrency trading bots that allows you to design a strategy with no prior knowledge. It connects directly to exchanges without the requirement for API permission. This software may be accessed through a tablet or smartphone.

Features:

- You can easily do technical analysis.
- Provides a drag-and-drop editor.
- This bitcoin trading bot is available 24 hours a day, seven days a week.
- It allows you to configure your plan with only a few mouse clicks.
- You may run an infinite number of backtests (the process of applying a strategy to historical data).
- Free of charge
- Apps for iOS and Android
- Binance.us, Binance, Bittrex, Liquid, Bitstamp, Kraken, HitBTC, Liquid, KuCoin, FTX are all Kryll.io exchanges.

Pros

- There is a mobile app that is simple, straightforward, and easy to use.
- Major exchanges are supported.
- Has a Demo

Cons

- There aren't as many exchanges supported as on other sites.
- Price structure is complicated.

Botsfolio

Botsfolio is a cryptocurrency exchange that allows you to trade for digital money easily. From a visible dashboard, you may track trade actions using this program.

Users

- Beginner traders with minimal trading experience.
- Traders looking for an automated trading system.

Features:

- It enables you to trade without any technical knowledge.
- You can use this tool to determine your investing objective.
- You can easily connect to your exchange.
- Allows you to trade for Bitcoin, Monero, Bitcoin Cash, and other cryptocurrencies.

Botsfolio Pricing

Botsfolio charges three pricing based on the value of your portfolio. The fees are paid in Ethereum, Bitcoin, Bitcoin Cash, and Litecoin.

- The price is $60 per year for a portfolio worth between $1,000 and $3,000.
- The price is $120 per year for portfolios worth more than $3k to $10k.
- The yearly fee is $540 for amounts ranging from $10,000 to $50,000.

It should be noted that all of the previous investments are subject to a 15% performance fee, which is levied quarterly on earnings generated only by the bot.

Pros

- No prior knowledge is required to begin.
- Intelligence for risk management.
- You have control over your investments.

Cons

- Only available on the Binance platform.
- A minimum investment of 1000 Tether is required.

NapBots

NapBots is a cloud-based method for using automated trading bots. It works with many cryptocurrency exchanges, including Bitmex, Binance, Okex, etc. This platform allows you to execute buy and sell orders automatically.

Users

- Traders in need of direction: rookies, specialists, and everyone in between
- Traders who are not constantly monitoring their accounts
- Traders that try to earn quick money by taking advantage of market fluctuations
- Cryptocurrency fanatics

Features:

- It provides a secure trade environment.
- More than 15 trading techniques are available.
- Allows you to manage your funds easily.
- You will be able to monitor the return on your investment using your cell phone.
- Using an API key, you can keep your money safe.
- The cost is $8.51.

- Binance, Kraken, Bitfinex, Bitmex, Bitstamp, OKEx, Bitpanda, and Phemex are all exchanges.

Pros

- Automation that aids in making wise investment decisions
- Site that is simple to comprehend and use
- Connectivity to many prominent exchanges
- In France, the AMF regulates the industry.

Cons

- Because it is located in Europe, there may be service delays.

Mudrex

Mudrex is one of the top crypto trading bots that easily construct a plan. It allows you to distinguish between excellent and bad investments, and this software lets you select between reward and risk based on your needs.

Features:

- It provides historical data for the testing approach.
- Mudrex accepts ByBit Coinbase Pro, OKEx, Binance, Deribit, and other cryptocurrencies.
- Provides an easy-to-use interface.
- It assists you in maximizing your ROI.
- Free of charge
- Binance, Coinbase Pro, OKEX, BitMEX, Deribit, and Bybit are examples of exchanges.

Pros

- Strategy builder with drag-and-drop functionality
- Backtesting and live trading in real-time

- A community that is active and involved
- Fees are low!

Cons

- There is no way to add custom indicators (yet)
- Mobile website functionalities are limited.

Haasonline

Haasonline is one of the greatest cryptocurrency trading sites, allowing you to trade on over 16 bots. It provides more than 50 technical indications, insurances, and safeguards that may be utilized to develop complicated strategies.

Features:

- It provides programmable script bots developed in C#.
- Backtesting may be integrated with the Discord (VoIP) and Telegram applications.
- This program has a dashboard that may be customized.
- It provides a lesson to help you grasp the platform.
- For three months, the price is 0.017 bitcoin.
- Binance, Binance Futures, Binance US, Bitfinex, BitMEX, Bitpanda Pro, Bitstamp, Bittrex, Bybit, CEX.IO, Coinbase Pro, Deribit, FTX, Gemini, HitBTC, Huobi, Ionomy, Kraken, KuCoin, OKEx, OKEx Futures, Poloniex.

Pros

- There are over 100 bots accessible.
- A large number of exchanges and trading pairs are provided.
- HaasScript allows for extensive customization.
- There are several indications, securities, and insurances.
- Privacy

Cons

- There is no free trial.
- Setup is more complicated than with other bots.
- One of the costliest bots

Wunderbit Trading

Wunderbit Trading is a cloud-based tool that allows you to invest in bitcoin easily. You may use this website to construct a fully automated trading bot.

Features:

- It allows you to choose the sort of assets you want to trade.
- This platform uses DCA (Dollar-Cost Averaging) to improve position entry price.
- It can keep track of your balance automatically.
- You can easily manage various cryptocurrency exchange accounts.
- Free of charge
- Android Mobile Apps.
- Binance, BitMEX, FTX, FTSUS, and HitBTC are all exchanges.

Pros

- The UI is simple and straightforward.
- Funds are secure due to API restrictions and 2FA.
- With Social Trading, you may take a "set it and forget it" attitude.
- The platform is entirely free.

Cons

- The platform is still in its early stages and does not have a significant user base.

- There is no mobile app for the platform.

The Gunbot

Gunbot is a customizable trading bot that works with over 100 different exchanges. It allows you to tailor your approach. Users may construct an endless number of bot instances with this automated bitcoin trading tool.

Features:

- One of the top automated crypto trading bots allows for automatic bitcoin trading for more consistent outcomes for your business.
- It offers a simple working technique that allows you to get started quickly.
- This bitcoin trading bot provides free updates without the need for a membership.
- This Binance trading bot allows you to earn from your trading methods.
- B 0.01375 as a starting point
- Binance, Bittrex, Binance US, Bithumb, Bitmex, Bitfinex, Bitmex Testnet, Coinbase Pro, HitBTC, Huobi Global, OKEx, OKCoin, Bitstamp, KuCoin, Poloniex, Cex, Kraken, Coinex are some of the exchanges available.

Pros

- 123 available swaps
- Permit for a lifetime
- Support for Windows, Mac, and Linux
- Excellent community support
- Extremely adaptable

Cons

- Available only on desktop

- For more seasoned traders

ProfitTrailer

ProfitTrailer is a tool that allows you to trade for crypto cash easily. It provides an easy-to-use dashboard with a comprehensive picture of your trade. This tool allows you to personalize your currency purchasing tactics.

Features:

- Allows you to view your sales without difficulty.
- You can look at potential transactions.
- It aids in the protection of your liquidity.
- You may purchase and trade numerous cryptocurrencies.
- The cost is $36.46
- Poloniex, Bittrex, Binance, BinanceUS, Binance Futures, Binance DEX, Kucoin, Huobi, BitMEX, and Bybit are all exchanges.

Pros

- A free version with paper trading is available.
- Multiple cryptocurrency exchanges are supported.
- A broad range of payment options is available.
- There is a large community behind these useful add-ons.

Cons

- There is no money-back guarantee.
- The subscription is quite costly.

Cryptotrader

Cryptotrader is an easy-to-use software for trading Bitcoin exchanges. This trading platform makes it simple to purchase and sell trading techniques.

Features:

- It is not necessary to install any software to utilize it.
- Provides a variety of tools for conducting a live test.
- Allows you to operate live trading bots with ease.
- This application offers customer service via email and SMS notifications.
- Free of charge
- Coinbase, Binance, Bitstamp, Bitfinex, Poloniex, Kraken, Bittrex, OKCoin, Huobi, and CEX.io are available exchanges.

Pros

- All popular Bitcoin exchanges are supported.
- Notifications are sent immediately.
- There are five distinct GUI models.
- Supposedly high hit rate
- Supposedly large potential ROI

Cons

- There is no assurance of huge earnings.
- There are monthly subscription costs.

HodlBot

HodlBot is one of the top bitcoin trading tools that easily design a personalized portfolio. This automatic cryptocurrency trading bot enables you to build a portfolio of over 350 currencies. This finest crypto bot program allows you to control your bot through a user-friendly interface.

Features:

- This Binance trading bot uses SHA 256 to encrypt your data.

- Allows you to rebalance your portfolio easily.
- Allows you to perform marketing indexing with less effort.
- This Binance bot trade can safeguard API credentials against unwanted access.
- This finest Crypto Trading Bot Platform keeps your portfolio on pace automatically.
- HodlBot Bitcoin bot may be tailored to your specific requirements.
- The monthly cost is $3.
- Binance, Kraken, KuCoin, and Bittrex are examples of exchanges.

Pros

- Simple to use
- Affordable
- Users keep 100% of their earnings.

Cons

- Not for experienced traders.
- Only for portfolio management and diversification.

6

EXTRAS

Differences Between Long and Short Positions in Crypto Trading

Long or short?

When it comes to trading, long positions are purchase orders placed by traders who hope to profit from an asset's rising price — in this case, cryptocurrencies. On the other hand, short positions are sell orders that are often made in negative markets.

While the notion of long and short is straightforward, knowing the fundamentals underlying long vs. short trading is critical for all traders.

Meaning of Short and Long Positions

Long and short bets represent the two alternative price directions necessary to profit. Long traders anticipate that the price will rise from a certain point, and those that go short expect the price to fall from their entry point.

Going long is about buying the cryptocurrency while going short has to do with selling the cryptocurrency. In a long position, the crypto trader has acquired a virtual currency and is waiting for its price to rise before selling. Some traders like to go long more often than they go short. Such traders are called bulls since they attempt to profit from bullish markets. This is true for the cryptocurrency industry, often on the rise.

When looking over longer time horizons, bitcoin is one of the most valuable digital currencies in market cap. A cryptocurrency's market capitalization (or market cap) measures its market worth.

It reached an all-time high at the start of 2021. This is also true for most digital currencies, which profit from overall optimism. As a result, short sellers frequently trade against the market. However, even if the cryptocurrency market is progressively increasing, it exhibits large drops following practically every gain, indicating that there is still plenty of opportunity for negative investors. Those who go short gamble against the coin and predict its price will fall.

Long and short positions are ubiquitous and apply to all markets and assets. It may also be used with hybrid investment products such as derivatives, which are tradable securities that track the price of an underlying asset. As a result, derivatives enable you to go long on an asset without really owning it. Futures exchanges, for example, allow you to profit from variations in the Bitcoin price without having to acquire or sell the cryptocurrency. Derivatives may be thought of as a type of gamble on the price of an asset.

Traders can combine long and short positions to construct dependable hedging methods that reduce possible losses.

Differences between Short and Long Positions

Short and long positions are diametrically opposed. A short trade on the same digital asset depletes the balance when a long transaction makes money. Furthermore, the mentality of bulls and bears differs in that bears are more cautious, while bulls want to take risks and experiment with new techniques. These, however, are simply stereotypes. During bull markets, short-sellers also take risks.

Understanding Short Position

Short trades, as previously stated, are the ones that create profits while betting against a coin. You typically purchase the corresponding currency when you sell a cryptocurrency, whether fiat or crypto. Thus, you purchase US dollars, USDT, or any other cryptocurrency or fiat money when you sell Bitcoin.

When traders predict the price of a digital currency to fall, they should go short. For example, if you are a day trader and believe that the price of Bitcoin will fall in the next days or weeks, you may choose to establish a short position. You may sell bitcoin for fiat currency and subsequently repurchase it at a lower price, or you can go short using futures, options, contracts for difference (CFDs), or other derivatives.

You should conduct a market analysis before making a selection. Day traders should use technical and fundamental analysis. Fundamental analysis is a method that cryptocurrency investors use to determine the intrinsic worth of a crypto asset.

After Setting a Short Position

Certain chart patterns and candlestick patterns predict a price correction or decline. Bearish chart patterns include the Double Top, Head and Shoulders, and Triple Tops. The Hanging Man, Shooting Star, and Gravestone Doji are bearish candles to look for.

When the price of a cryptocurrency falls below a stable support level, some traders opt to go short. This is especially true if the price has been trading for a period within a channel, whether bullish, bearish, or horizontal.

Alternatively, you might go short by taking advantage of price fluctuations inside the channel. As a result, you won't have to wait for a breakout. Instead, if the price is approaching the resistance of a channel, you might go short in the belief that it will return to the support level.

Regardless of the type of study you select, you must be certain that the price will fall if you intend to establish a short position. You will end up trading against the market if you do not.

Understanding Long Trades

Traders may be interested in going long if the market is bullish. Opening long positions during rallies, such as the one we witnessed at the end of 2017 or the current long-term rise that began in November of 2020, may bring significant profits. As previously said, you should ensure that your decision is supported by accurate market analysis. You may either wait for the price to break over a high resistance level or go long during the current rally hoping that it will continue for a while.

Another strategy used by affluent investors is to acquire and hold the coin. In this situation, they are not actively trading it but rather keeping it for months or even years with the idea that the cryptocurrency's value would rise despite mistakes and adjustments.

This technique necessitates significant investments and is not ideal for retail traders looking to speculate on Bitcoin and altcoin volatility in the short term. On a related issue, huge traders contradict market patterns by buying the dip and selling the rip, which means they go long when the price is correcting.

Day traders may wait for the price to reach the oversold level of the Relative Strength Index (RSI) or Stochastic RSI to go long. Alternatively, purchase traders can employ chart patterns like the Double Bottom, Inverse Head, Shoulders, Ascending Triangle, or Hammer. Bullish candlestick patterns also exist, such as the Hammer and Dragonfly Doji.

After Opening a Long Position

Keep in mind that, in comparison to foreign exchange or equities, the cryptocurrency market is in its infancy. As a result, technical analysis may not be as effective, as the price might frequently catch you off guard. Nonetheless, you should evaluate all aspects that might influence the pattern.

The good news for optimistic traders is that digital currencies behave similarly to corporate stocks, unlike forex pairs, which have no long-term aim. They are mostly traded against the US dollar and tend to rise in value.

Going Long or Short in Financial Markets

Yes. Traders take long and short positions in all markets. Indeed, this is the basic essence of trading. Therefore, you can't do anything other, regardless of the market. There are more sophisticated variants of long and short if you trade derivatives. For example, if you trade options, you will utilize a variety of long and short combinations.

Option contracts provide holders the right (but not the responsibility) to purchase or sell an underlying asset at a defined strike price on a certain date or before it. They are widely used in all financial markets, and traders often employ call options to go long and put options to go short.

They can, however, combine these. Traders, for example, can short both call and put positions. If you sell a call option, you are betting against the result of that particular call option, which is bullish. As a result, you are pessimistic about the underlying asset's price. You may go long on both calls and put options in the same way.

Takeaway

Long and short positions are at the heart of trading, and traders attempt to identify emerging patterns to make sound judgments. However, deciding whether to go long or short is only half the battle; determining the optimum entry point is equally critical.

Last but not least, certain cryptocurrency exchanges provide margin trading, which may significantly increase the goals of long and short positions. However, you should know that the risk is greater than when trading with your cash while using leverage.

Top Cryptocurrency Trading Mistakes

If you have money and want to undertake a high-yielding but high-risk investment before you rush in, you should be aware that it is a highly volatile digital asset with equal odds of making a fortune or going bankrupt. The mistakes listed below might help you better understand the Cryptocurrency market and avoid repeating them.

- Investing Without Understanding

Investing in digital money or another asset before fully understanding the details is the most stupid move one can make. Before you invest even a single rupee, you should know what you're getting into. You should gather information about how it works to act wisely in any unfortunate situation.

- Encryption is associated with security.

You read that properly. Cryptocurrencies are encrypted, but to make them more secure, this does not guarantee they cannot be hacked or stolen. Because these resources are decentralized, you are alone responsible for their protection.

- Failure to Pay Attention to the Math

The goal of investing is to maximize profit potential. Because it has been predicted that bitcoin will rise in value in 2021, you should keep your eye on the broader picture. Furthermore, how would you go about doing so? Focusing on the numbers will reveal whether or not you are generating again.

You should look into transaction fees. Furthermore, because cryptocurrencies may be pretty volatile, several price changes will be in a day or even 60 minutes. As a result, if you need to take advantage of these developments, you should look at exchange rates.

- HODL

HODL indicates keeping your investment regardless of how volatile the market is. However, you may discover that you do not have the opportunity to wait for a substantial reward from your investment at times. That is the moment to get rid of your troubles.

- FOMO

FOMO, or Fear of Missing Out, refers to investing in promotion.

- FUD

FUD stands for Fear, Uncertainty, and Doubt. As obvious as it appears, FUD may prohibit you from investing in cryptocurrency regardless of if the inspection facts or market sentiments back you and encourage you to contribute.

- Investing in Only One Cryptocurrency

Bitcoin may be approaching different folds, but no one knows when the market will turn back. As a result, it is prudent to invest in more than one cryptocurrency, as certain distinct monetary standards, such as Ethereum and Altcoin, may provide excellent returns. An adage in donating says, "Don't lock up your funds in one spot." Follow that advice, and don't put all of your money into a single cryptocurrency.

- Getting Intimidated by Market Volatility

Crypto is a highly volatile market with frequent fluctuations. This is enough to frighten any novice gamer. While prepared financial backers are used to the nature of a crypto market, they may bag benefits in any event amid value crashes based on their knowledge and expertise. Fledglings will be in a frenzy, and frenzy is maybe the finest rival on the market. It may induce you to appreciate panic selling, which occurs when financial supporters sell their assets on the basis that the market has reached the absolute bottom.

- Not Considering the Stop Loss Method

While suppressing your emotions can be a difficult game to play for many financial backers and may require some work to dominate, one skill that any dealer or financial backer should have is the ability to recognize the misfortune and proceed. Setting up a stop-loss plan requires selecting how much you will lose before speculating.

- Using Low-Interest Rates as a "Take Deal"

There is no way to admit that people are supporters of the "sell" attitude. Costs in the cryptocurrency market are typically low, which is to be expected. As a result, before investing in a coin, you should thoroughly investigate why it is valued so low.

- Using the Wrong Exchange

Choosing the wrong exchange might have long-term consequences for your trading experience. If you utilize a sham exchange, you risk losing your whole crypto holdings, in addition to losing your earnings. As a result, you must check that the exchange in which you will deposit your money is safe, reliable, and secure.

- Absence of an Exit Strategy

Even after making a significant profit, most cryptocurrency traders do not become aware of how to plan their exit strategy. This is the point in a transaction: you either grab your profit or halt your losses.

Remember that cryptocurrency is not a bank, and its value can fluctuate at any time. To be a good trader, you must first understand protecting your winnings. You may use the stop-loss tool, accessible on major exchanges, to reduce your losses while implementing an exit plan. The stop-loss is a trading instrument used to restrict a trade's maximum loss by automatically selling assets when the market price hits a certain level.

Also, don't get caught up in FUD (fear, uncertainty, doubt) or greed when profit-taking. Be confident in your ability to depart after you have earned a profit, which, as previously said, is much more likely if you have done your study. Another piece of advice for exiting your earnings is to take a profit at a predetermined period. You may, for example, elect to sell 20% of your BTC earnings every time the price doubles.

- Forgotten Password / Key

While only 21 million bitcoins will ever be created, fewer will be accessible for trading since many of them will be lost forever because users have forgotten the passwords to their digital wallets.

You can't always contact someone to change your password; you're locked out if you forget or lose it. According to Chainalysis, a cryptocurrency research business, over 20% of Bitcoins produced have been lost in stranded wallets. As a result, how you save your password is crucial and should be considered before you begin trading.

- Not Extremely Careful

Many beginners and experienced cryptocurrency traders, believe it or not, have lost money by sending their coins to the wrong addresses. Unlike a bank wire transfer, which can be stopped, or a check, which can be canceled, there is frequently no recourse if you make a fat-finger mistake.

When conducting a transaction, it is critical to check an address numerous times, twice or three times. Someone, for example, can simply transmit USDT to a USDC address. A bitcoin address is a string of alphanumeric characters representing the address of a wallet, exchange, or other blockchain-specific address. All wallet and exchange addresses are unique and indicate the sender and receiver's position on the blockchain network.

- Purchasing More Than You Can Afford to Loose

One common pitfall for newcomers is investing money they need to fulfill their requirements, such as rent or mortgage payments, into cryptocurrency. Investing money you can't afford to lose might get you in hot water if the coin's value plummets after you acquire it.

Of course, you never know when coins may appreciate or depreciate, so it is better to play it safe by investing only money that you can afford to lose in the worst-case situation.

- Purchasing the Wrong Coin

Buying the wrong coin is one of the most common blunders that beginning investors make. It's not Bitcoin just because it contains the word "Bitcoin" in its name. Here are several examples of coins that have the word "Bitcoin":

Bitcoin Gold

Bitcoin Cash

Bitcoin SV

Bitcoin Private and dozens of other cryptocurrencies with the term "Bitcoin" in their titles are direct offshoots of the original bitcoin.

That is not to argue that these offshoots are nasty or fraudulent; it is only that they are not the original bitcoin, which has been extensively traded and quoted. However, if you purchase the incorrect coin, it is not the end of the world. You may always sell it and acquire the correct one for a profit. Purchasing because it is inexpensive You don't want to buy coins because they're dirt cheap without doing your homework beforehand.

Consider the following additional factors:

- What is the function of the coin?
- What is the maximum available supply of coins?
- Understanding the Market Cap (number of coins available) can influence how valuable the coin can become in the future.

Top Cryptocurrency Tips

- Develop a crypto trading plan.

It is difficult to distinguish between real cryptocurrency advice and frauds; many sharks are out there eager to grab your money. In the first nine months of 2021, there were 7,118 reports of crypto investment frauds. According to Action Fraud, this was a 30% increase over the entire year of 2020, with the average loss per victim being £20,500.

Take a step back from the crowd when you're presented with a lot of information about a cryptocurrency. Examine the project critically. How many people utilize it? What issue does it address? Is there any connection to the industry? Avoid coins that promise the world but provide nothing concrete.

- Risk management

Some persons who provide cryptocurrency trading advice may not have your best interests at heart. Don't be stung by repeating the same mistakes as others. Set limitations on how much you invest in a certain digital currency and avoid trading with more money than you can afford to lose.

- Diversify your cryptocurrency portfolio.

It's not a good idea to put all of your money into a single cryptocurrency. Spread your money across multiple digital currencies in the same way as you would with equities and shares. This means you won't be over-exposed if one loses value, which is especially important given how unpredictable the market prices for these assets are.

There are dozens to select from, so do your homework. Worldcoin and Safemoon are two examples.

- Commit to it for the long term.

Prices can fluctuate substantially from day to day, and inexperienced traders are sometimes fooled into panic selling when prices are low. Cryptocurrencies aren't going away, and putting your money in the market for months or years at a period may provide the most satisfactory results.

- Make purchases more automated.

Just like with ordinary stocks and shares, automating your crypto purchases can help you take advantage of pound cost averaging. You can set up regular purchases on most cryptocurrency exchanges, including Coinbase and Gemini.

This is when crypto investors instruct the platform to buy a set quantity of their favorite cryptocurrency — for example, £100 worth of bitcoin – every month. It implies they receive somewhat less cash when prices are high and slightly more when prices are low.

You alleviate the burden of time on the market by either purchasing or selling a currency at what you believe is the lowest feasible price. Even market specialists have difficulty getting it properly.

- Make use of trading bots

Trading bots can be beneficial in specific situations, but they are not suggested for novices searching for cryptocurrency investing advice. They are frequently disguised scams. If a true algorithm existed that perfectly timed your buy and sell deals, everyone would be using it!

- Have a reason for participating in each deal.

While this may seem apparent, you must have a clear goal in mind while investing in bitcoin. Whether you want to day trade or scalp, you must have a reason for starting to trade cryptocurrencies. Trading digital currencies is a zero-sum game; you must understand that there is an equal and opposite loss for every success: Someone wins, and someone loses.

The cryptocurrency market is dominated by enormous 'whales,' similar to those that deposit thousands of Bitcoins in market order books. Can you guess what these whales excel at? They have patience; they wait for inexperienced traders like you and me to make a single mistake that results in our money ending up in their hands owing to preventable errors.

Whether you are a day trader or a scalper, it is sometimes preferable to not gain anything on a particular deal than to hurry your way into losses. Based on our years of market monitoring, we can confidently inform you that you can only remain successful by avoiding specific trades on some days or times.

- Set profit objectives and apply stop losses.

If you've never heard of the word "stop-loss" in trading, go on this link to learn more about it. Every transaction we enter necessitates knowing when to exit, regardless of whether we make a bitcoin profit or not. Establishing a precise stop loss level may help you limit your losses, which is a skill that most traders lack.

Choosing a stop loss is not a random activity, and maybe the most important thing to remember here is that your emotions should not carry you away. A good place to set your stop-loss is at the cost of your coin. Set the least point you're prepared to swap your currency at if you bought it for $1,000, for example. This ensures that if the worst happens, you will be able to walk away with the money you invested in the first place.

The same is true for profit levels; if you want to exit the market after making a particular minimal profit, adhere to it. Don't be greedy; it never looks good on anyone!

Mastering the world of cryptocurrencies will take time, so don't put too much pressure on yourself to become an expert on day one. It takes time, effort, and ongoing learning to figure out how to maximize the potential of your virtual currency, whether you want to utilize them for trading or everyday transactions.

CONCLUSION

B uying cryptocurrency from an exchange was straightforward in the early days of cryptocurrency. Individuals would create an account, log in, purchase bitcoins, and transfer them to a wallet. Recently, however, cryptocurrencies have been notable for their extreme volatility, with values moving wildly in minutes. Also, traders may participate in cryptocurrency trading from anywhere in the world and at any time of day. Slowdowns in exchanges and transaction delays aggravate the matter even further. However, investors do not have the time to devote to the cryptocurrency markets as much as is required to make the finest trades consistently. These variables, when taken together, restrict the efficacy of human cryptocurrency trading in some ways.

Similarly, when you try to make an order on a major cryptocurrency exchange these days, you will notice that another order has been made directly above it at a slightly higher price. You will also have to pay a somewhat greater price as a result. However, in reality, someone is not always trading in front of the computer. A crypto trading bot initiated the offer and decided to purchase a little higher in a fraction of a second. Crypto trading bots have taken over the whole industry. They're more effective than humans, particularly when executing trades. These trading bots conduct around 75% of all trades.

Trading bots are software programs that execute cryptocurrency's timely trades to generate profit. Ideally, the bots make a bigger profit in risk-adjusted terms than if you had just purchased and held the same cryptocurrencies. Traders utilize bots to take advantage of the global cryptocurrency markets, which are open 24 hours a day, seven days a week.

Bots have the edge over investors in that they can react faster. Most investors lack time to devote to constantly getting the greatest transaction, which bots can achieve. When it comes to crypto trading bots, you have the option of making riches or losing money on the financial market. Trading relies on many factors, including better strategies and determining the best time to trade. However, because transactions may go in any direction, there are no assurances that a strategy will succeed all of the time. To sum up, traders can benefit handsomely from crypto trading bots.

A MESSAGE FROM THE AUTHOR,

AS INDEPENDENT AUTHORS IT'S REALLY DIFFICULT TO GET REVIEWS.

THEREFORE, I'D REALLY APPRECIATE IF YOU PLEASE DO LEAVE A REVIEW OF THIS BOOK ON THE PLATFORM WHERE YOU BOUGHT IT.

MANY THANKS.

www.ingramcontent.com/pod-product-compliance
Lightning Source LLC
LaVergne TN
LVHW032255060326
832902LV00024B/4600